Coming Home to the Pleistocene

Paul Shepard

Edited by Florence R. Shepard

ISLAND PRESS / Shearwater Books
Washington, D.C. • Covelo, California

A Shearwater Book
published by Island Press

LIBRARY OF CONGRESS CATALOGING-IN-PUBLICATION DATA
Shepard, Paul, 1925–1996
Coming home to the Pleistocene / Paul Shepard ; edited by Florence R. Shepard.
p. cm.
"A Shearwater book"—
Includes bibliographical references and index.
ISBN 1–55963–589–4. — ISBN 1–55963–590–8
1. Hunting and gathering societies. 2. Sociobiology. 3. Nature and nurture. I. Shepard, Florence R. II. Title.
GN388.S5 1998 98-8074
306.3'64—dc21 CIP

Printed on recycled, acid-free paper

Manufactured in the United States of America
10 9 8 7 6 5 4 3 2 1

Contents

Preface

PAUL SHEPARD PROBABLY BEGAN this book when he was a child, following Ben, an older boy he idolized, as they set and checked traplines and hunted and fished in the Missouri woods near their homes. Paul's father was a horticulturist and director of the Missouri State Experimental Farm. Ben's father helped with the care of the experimental orchards and vineyards. The families lived atop a hill that overlooked woods and farms and the town of Mountain Grove. Paul, surrounded by a rich natural environment and the love of multiple caregivers, wandered freely as a child through the countryside. The excitement of his primal experiences of hunting and fishing as well as his near idyllic childhood nestled in his memories until his death and, I believe, formed a fundamental core of experience: the basis for his conceptualization of *Coming Home to the Pleistocene.*

I was committed in the editing of *Coming Home* to be true to Paul's ideas and present them as clearly as possible, retaining his distinctive voice throughout. As it turned out, I needed to do very little writing. Paul was a circular thinker, even more so perhaps toward the end of his life. He began with a premise that he worked and reworked from various perspectives, digging deeper and deeper, uncovering the radical center of things, sentence by sentence, chapter by chapter. I had only to search and find, embedded in the matrix of the text, the necessary introductions, explanations, and transitions. Positioned a bit differently, his own thoughts brought the clarity needed. In a couple of instances, I added an

Editor's Note from material that I thought he might have found pertinent. Otherwise I did not expand or alter his original ideas, though I have provided occasional transitions or clarifications. In these instances, I studied his notes and previous manuscripts as well as original sources to better understand his position. References have been difficult to track down in some instances. I have tried my best to verify all references and quotes but confess that there may be occasional lapses for which I apologize in advance.

Although she had no evidence that I would be up to the task, Barbara Dean, associate editor at Island Press, supported my desire to edit this book. All of the compliments bestowed upon editors, which, alas, sound so trite when written down, are absolutely true with respect to Barbara Dean: I was able to complete the task only because of her careful guidance and support. I was impressed throughout with her abilities and insights; her suggestions and hard questions I took seriously. She was understanding of my feelings and patient with me in the face of my own impatience. I shall always value the experience of working with her as one of the best in my lifetime.

I am deeply grateful, too, to Professor Emeritus John Cobb Jr. of the Claremont School of Theology, who critiqued the first chapter and provided helpful suggestions. I thank Barbara Youngblood and Christine McGowan, developmental and production editors at Island Press, and Don Yoder, copy editor, who provided the expertise needed to bring the manuscript to publication. My daughters Lisi Krall, Ph.D., and Kathryn Morton, M.D., were helpful throughout. Lisi organized Paul's references and did research when Paul was unable to get to the library. Kathryn, monitoring Paul's illness from start to finish, seeing that he received the best care and treatment possible, helped to keep him strong and able to complete the manuscript. Paul was at the center of a loving group of family and friends throughout the writing of this book. I am deeply grateful for their friendship and love. He was the light at the center of our fire circle.

Paul wrote the Introduction and put the finishing touches on *Coming Home* three weeks before he died. Once that was done and I had sent the manuscript off to Island Press, he succumbed peacefully and with great courage and dignity to the inevitable conclusion of his life. The book provided closure for a life lived with great creativity, joy, and love—a life committed to a vision of how we can become more fully human.

I began editing this book in early summer when the snow still glistened on the mountains and the Hoback Basin was emerald green and filled with birdsong. I have completed it in the fall as snow builds once more and the basin is tinted a bland beige-brown. Except for occasional calls from ravens flying over or coyotes circling at dusk and dawn, silence reigns. It has been solitary and lonely work, much of it done through the mist of mourning and remembering. But it has been "good work"—work that has brought me closer than ever before to Paul's poetic vision.

Florence R. Shepard
The Hoback Basin
Bondurant, Wyoming
October 1997

Introduction

THIS BOOK IS ABOUT our self-consciousness as individuals and our worldview as a species based on the biological legacy and cultural influences we inherited from our ancestors, the Pleistocene hunter/gatherers (also called foragers). We began our path to the present on ancient savannas where we vied for our lives with other predators that shared this earth with us. Our humanity evolved increasingly as we were able to see ourselves reflected in nature and in kinship with other species in the circle of life and death, a way of life in which all things living and nonliving were imbued with spirit and consciousness. In that archaic past we perfected not only the obligations and skills of gathering and killing, but also the knowledge of social roles based on age and sex, celebration and thanksgiving, leisure and work, childrearing, the ethos of life as a gift, and a meaningful cosmos. In this book we shall look into the unique mind of our hunter/gatherer ancestors as a way of understanding the wholeness of all that we think of as "culture" on this planet that we call home.

In a society committed to goals of development and progress, looking back is seen as regressive. Insofar as the past is seen as limiting, the modern temper has never been sympathetic to genetic or essentialist excursions into the complex processes of becoming and being human in the sense of prior biological or psychological constraints. Such appeals to atavism seem both illusory and antisocial at a time when the individual and the culture are regarded as socially constructed. Those who seek solutions to contemporary problems in the past—naturalists, ecologists, rural visionaries—must bear the labels of "regressive," "romantic," or "nostalgic."

Historical as well as ideological reasons work against reclaiming our human and prehuman past. The uncritical attribution of all good things to lost origins, an ignorant beatification of everything savage or primal, the misunderstandings of biological evolution as in, for instance, social Darwinism, and the lack of fully understanding the importance of ceremony and myth in personal and social processes—all have contributed to a yearning for lost paradises and the search for vanished paragons. These distortions of the truth of our past have subjected dialogue on "the uses of the past" to ridicule, a fate to which my own efforts to reconceptualize our primal forebears fell victim.

In the early 1970s, after two decades of activism, and after publishing my first book, *Man in the Landscape*,[1] I became disillusioned with the environmental movement. More to the point, I no longer believed that understanding the meaning of ecology would make any difference in turning the public's consumptive mind to a more sustainable economy. In 1972, I had brought to Scribner's attention José Ortega y Gasset's *Meditations on Hunting*,[2] for which I had written an introduction and found a Hispanist translator, Howard B. Wescott. Reviewing the new anthropological information on hunting/gathering peoples, I then tried to detail the claims of the past upon the present in a book of my own. In 1973 when I published my first book on the world of hunters and gatherers, *The Tender Carnivore and the Sacred Game*,[3] it was not received as "good news." I expected as much; reviewers found it easy prey. It soon went out of print with the minor distinction of having become a curiosity and a marginal cult object.

I was, of course, not the only one to try to formulate the meaning of hunting and gathering for our own time. Even so, few efforts were made by mainstream scholars to sort out the significance of the lives of hunters and gatherers. The eagle eye of the humanist and his modern educated counterpart were always scanning for romantic nonsense. Even sympathetic writers pretended that hunting signified at best only a lost past.

Everything I have written since that time was influenced by what I uncovered in my research on *The Tender Carnivore*: our perception of animals as the language of nature in *Thinking Animals* and *The Others;* the "natural" way of childrearing in *Nature and Madness;* and the bear as a dominant sacred animal connecting people ceremonially to the earth in *The Sacred Paw.*[4] Recently I have returned to the theme of our hunter/gatherer ancestry in presented papers and published essays. "A Post-Historic Primitivism" was first delivered at an interdisciplinary conference on wilderness and civilization held in 1989 in Estes Park, Colorado, and later published in an anthology, *The Wilderness Condition,*[5] growing out of that conference and edited by Max Oelschlaeger. In 1993 I presented a paper, "Wilderness Is Where My Genome Lives," at the International Conference on Wilderness at Tromsö, Norway, that was later published in *Whole Terrain.*[6] These essays were expanded into the framework for this book. Through writing and contemplation over the years, I have somehow bonded firmly to those ancient ancestors, their society and ecology, and this kinship has guided my writing and thinking.

During the past twenty years new information on Paleolithic peoples has emerged: analysis of prehistoric art, the lifeways of present-day hunter/gatherers, the bio-ecology of hunting/gathering, the psychological and cultural dynamics of myth and ceremony among tribal peoples, the origin of other basic economies—especially agriculture and pastorality—and the role of genes in human behavior and health. Much that was speculative in 1973 has been strongly supported by new evidence showing "primitive" or "ethnic" peoples to be as complex, profoundly religious, creative, socially and politically astute, and ecologically knowledgeable as ourselves, or more so, and at the same time to be equally subject to individual human frailty and to aggression, lying, stealing, and cheating. In the interval, numerous anthropologists have published work on nonliterate, tribal peoples that justifies our attention to and regard for their lifeways.

✦

IN THIS BOOK I have touched upon some questions that have persisted in my mind since the writing of *The Tender Carnivore*. If a human way of hunting and gathering is replaced by agriculture and village life, what does agriculture advance and what does it lose? Why did it come into existence? Rural life has a disarming appeal that is also fashionably connected to a form of feminism and to the resurrection of the goddess. What does this have to do with pastorality? How do the two agricultures that replaced foraging—farming and pastorality—deal with the cycle of life and death and why is death central to the discussion? There is a dialogue between "the wild" and "the domestic" to be understood here. How can we understand that dialogue in terms of their metaphysics? Does culture really *evolve* and is its "evolution" inevitable? Irreversible? What could reversibility mean?

As I complete this book, I see new questions that deserve consideration in the future: How is one to accommodate an ethics of normal killing—the mien of the predatory human—and the ethics of widespread infanticide by mothers? In addition to preying upon them how can one pray *to* animals? Is esthetics an adequate instrument for contemplating the huge body of painting on rock and sculpting of bone and antler? Can we even contemplate the good life without institutional Great Art and Classical Music, museums, theater performance, a written legal code, awareness of other cultures, armies, written history, a moral basis of community based on the Greek city, information flow, economics of industrial distribution and storage, and advanced medical and technical protection from disease and weather? Given its possibilities rather than its reality, is the city something we can give up? Or must our modern amenities be sacrificed for us to become savage again?

The literature of environmentalism has descended on the Western world like a pall during the past quarter-century, so it is not surprising that many people find the constant review of environmental destruction and species extinction too much to bear. At their most incisive the "cures" address not only our fundamental beliefs but civilization itself. And we are so imbued with the virtues of civility—the high moral ground of ethics and social community—that all we hold dear seems threatened by any suggestion of an atavistic regression to our natural selves.

We have placed rural and urban life in opposition when, in fact, the two are one—part of the same dream of a subjugated natural world transcended by the human spirit. From its beginning agriculture made the village and then the city possible. And the city continues to depend not only on the material production of farms and ranches, but also on the social practices that create an explosive demography that feeds the corporate and industrial exploitation of the earth.

We perceive the dark side of our present condition as our failure to adhere to the standards of "civilization." Crime, tyranny, psychopathology, addiction, poverty, malnutrition, starvation, war, terrorism, and other forms of social disintegration seem to be the weaknesses and flaws in our ability to live up to the expectation of being civilized. Present disillusion with the ideologies and goals of "advanced" nations since the Enlightenment, and the decline in quality and experience of life itself, are matched by the degradation of world ecosystems and the ratcheting scale of poverty and widespread social turmoil. In the absence of some new synthesis that rejoins us to our natural heritage, the world of corporate organization pushes us toward the degenerating process of conformity, the frenzied outbreak of genetic engineering, and the pied piper's technological tootle leading down the "information highway" toward the "networked" insanity that confuses electronic regurgitation with wisdom. This circuit-sedative turns us into entertainment junkies hooked without reprieve to the economic machine and its media, a new level of confusion between reality and virtual reality. Our image of ourselves—of humanity—is in question because ideology alone always fails. Species and cultures that have endured for scores of thousands of years are subject to oblivion in the hands of this culture in which our faith has been upstaged by growth.

We are not new as organisms or as a species, nor are the millions of species of plants and animals around us new. Somehow our hunger for change and novelty has cost us a sense of the role of nature in personal growth and the necessity of compliance and limitation. We must now ask in what sense our present dilemmas are measured by departure from some kind of diffuse, primordial scheme of human life and what is possible in terms of recovery.

In the face of predominant anthropocentric values, the vision of *natural* humankind seems eccentric, regressive, even perverse. Our idea of ourselves embedded in the context of the shibboleth of growth places us at odds with the notion of kinship with nature. When we grasp fully that the best expressions of our humanity were not invented by civilization but by cultures that preceded it, that the natural world is not only a set of constraints but of contexts within which we can more fully realize our dreams, we will be on the way to a long overdue reconciliation between opposites that are of our own making. The tools we have invented for communicating our ideas and carrying information have actually impaired our memories. We must begin by remembering beyond history.

Notes

1. Paul Shepard, *Man in the Landscape: A Historic View of the Esthetics of Nature* (College Station: Texas A&M University Press, 1991); first published in 1967.

2. José Ortega y Gasset, *Meditations on Hunting*, trans. Howard B. Wescott, Introduction by Paul Shepard (New York: Scribner's, 1972).

3. Paul Shepard, *The Tender Carnivore and the Sacred Game* (New York: Scribner's, 1973).

4. Paul Shepard, *Thinking Animals: Animals and the Development of Human Intelligence* (New York: Viking, 1978); *Nature and Madness* (San Francisco: Sierra Club Books, 1982); *The Sacred Paw* (New York: Viking Penguin, 1985); *The Others, How Animals Made Us Human* (Washington, D.C.: Island Press, 1996).

5. Paul Shepard, "A Post-Historic Primitivism," in Max Oelschlaeger, ed., *The Wilderness Condition: Essays on Environment and Civilization* (San Francisco: Sierra Club Books, 1992).

6. Paul Shepard, "Wilderness Is Where My Genome Lives," *Whole Terrain* (1995–1996): 12–16.

I

The Relevance of the Past

History is not a chronicle but a Hebrew invention about the way the cosmos works, a notion that became the accepted "word" for the civilized world. One of the problems with this version is that it does not see the past reoccurring in the present. Yet Octavio Paz reminds us: "The past reappears because it is a hidden present. I am speaking of the real past, which is not the same as 'what took place.' . . . What took place is indeed the past, yet there is something that . . . takes place but does not wholly recede into the past, a constantly returning present."[1] History as written documentation of "what happened" is antithetical to a "constantly returning present," and as a result its perception of time and change is narrowly out of harmony with the natural world. Written history is the word. Time is an unfinished, extemporaneous narrative.

Prehistoric humans, in contrast, were autochthonous, that is, "native to their place." They possessed a detailed knowledge that was passed on from generation to generation by oral tradition through myths—stories that framed their beliefs in the context of ancestors and the landscape of the natural world. They lived within a "sacred geography" that consisted of a complex knowledge of place, terrain, and plants and animals embedded in a phenology of seasonal cycles. But they were also close to the earth in a spiritual sense, joined in an intricate configuration of sacred associations with the spirit of place within their landscape. Time and space as well as animals–humans–gods–all life and nonliving matter formed a continuum that related to themes of fertility and death and the sacredness of all things.[2] During prehistory, which is most of the time that humans have been on earth, the dead and their burial places were venerated and mythic ancestors were part of the living present, the dreamtime ones whose world was also the ground of present being. Ignore them as we will, they are with us still.

The roots of history as written, as Herbert Schneidau has shown us,[3] were formulated by the Hebrew demythologizers who created a reality outside the rhythmic cosmos of the gentiles who surrounded them and who were grounded in prehistoric, mythical consciousness with rituals of eternal return, mimetic conveyance of values and ideas, the central metaphor of nature as culture, and, most of all, the incorporation of the past into the present. Unlike history, prehistory does not participate in the dichotomy that divides experience into good and evil, eternal and temporal. Rather, it belongs to a syncretic system that accepts multiple truths and meanings and attempts to reconcile them. This state of consciousness is not due to a rational process. The mythic mind, as John Cobb Jr. has explained it, does not recognize the "separateness of subject and object" but instead sees "a flow of subjective and objective contributions . . . bound together" where there is no "clear consciousness of subject as subject or of object as object."[4]

The Hebrews, who initiated the move away from the earth and toward the historical view, did not try to reconcile opposing beliefs. Nor did they have a sense of place. To the contrary they insisted on "deracination from the spirit of place" and asserted that they were "journeyers" to the Promised Land.[5] In the Hebrew view, the realm of the sacred was granted only to Yahweh. Objections to the oldest traditions of time and the past imply a deeper strain that has to do not with the content of history, but with a self-conscious alienation that first became evident in the Hebrews. The assault on the local wisdom of primal peoples culminated in the outwardness of nature and the inwardness of the personality.

Focusing on heavenly domination over earthly phenomena, history became an attempt to look away from earth. The Hebrews and the

Greeks, who were their contemporaries and whose parallel culture in the Eastern Mediterranean shares many common features with that of the Hebrews,[6] saw alienation as the touchstone of humankind. They understood themselves as outside the nature-centered belief systems of other peoples, whose cosmologies linked past, present, and future in stories and art with eternal cycles and sacred places. History is a way of perceiving human existence that opposes and destroys its predecessor, the mythic world, which sees time as a continuous return and space as sacred, where all life is autochthonous.

The Hebrew and Greek founders of history were not so intent on rejecting nature as they were on understanding temporal events as unique. The Hebrews, the Greeks, and, following in their steps, the Christians asserted that events are novel, uncertain, tangential, and contingent rather than embedded and structured, the result of the thoughts of a living, omniscient, unknowable God. The past was a highway on which there could be no return.

The prototype of this linear sequence of ever-new events, where nothing was repeated and to which nothing returned, was the *Old Testament*, a record of tribal endogamy, identity, and vision. Thirty-two hundred years later, history has grown fat with the civilized written records that replaced oral traditions and added vast secular data to religious history. This breakaway from the mythic life, which linked our species to the natural world, began when the early Hebrews rejected the nature/process stories and rites of their pagan contemporaries for the myth of a single god who, outside the world, reached into his creation, willfully deranging its rhythms, acting arbitrarily, making life a kind of novel, a history. The effort of the Hebrews to distance themselves from the sacred immanence in the natural order initiated what Cobb has called the "reflective consciousness" of humankind in the "Axial Period" (between 800 and 200 B.C.)—a transitional state of human cognition in which the archaic mind was altered and a "conscious control of symbolization and action" emerged.[7] Although the Hebrews had begun to develop a "reflective consciousness," a state of consciousness in which they actively attempted to understand their place in the greater scheme of things, they were still locked into a kind of projection of their unconscious that symbolized subjective elements arising from a deep substratum of the mind.

The Greeks, Cobb argues, succeeded in distancing themselves from sacred immanence in the natural order through further development of the "reflective consciousness." The change of the "structure of existence," the way they envisioned the possibilities in their lives, moved from the unconscious to the conscious. In the case of Homeric Man, "the object of conscious experience . . . was primordially the sensuously given world . . . in which the subjective was subordinated to the objective." Out of this grew "esthetic distancing"—an ability to see beauty in the world, in nature, in the human body, and in temples and other human artifacts, esthetically pleasing forms corresponding to rational psychic structures. By esthetic projection of beauty and perfection onto their gods, Cobb says, "the Greeks subordinated mythical meanings to the rational consciousness. . . . Gods were conceived as visual objects" and an "intelligible order" was imposed on the myths. In this way the mythical became the mythological. Things could be treasured for their beauty as opposed to their utility or their numinousness. Careful study of the objects of art resulted in "demonstrated laws of form and quantitative mathematical laws,"[8] which allowed replication and thus the development of mathematics, natural science, philosophy, drama, and performance music. Science and esthetics emerged together—invented for the West, so to speak, by the Greeks. Greatness was equivalent to excellence and beauty rather than to morality. The gods became drama and sculpture; nature was reduced to the sensed source of intellectual description and artistic power.

For Christians the crucial events "on earth" were finished except for a final judgment. Christian existence was defined as spiritual existence that expressed itself through "radical responsibility for oneself" as well as "self-transcendence" through love of others.[9] Christians further emphasized the distinction between the word of a patriarchal god and all myths of an earth mother—thereby separating themselves even more from the numinous earth and its processes.[10] Individual responsibility for self-scrutiny in terms of sin or good works took precedence over the timeless sacredness of the earth and its processes. The notion of the unreturning arrow of historical time in the Western mind was taken up by Christianity.

✦

OUR HUNGER FOR HISTORY, our obsession with it, is exacerbated by the lack of meaning in our own personal experience created by the historical attitude. Herbert J. Muller presents us with a paradox: "Our age is notorious for its want of piety or sense of the past. . . . Our age is nevertheless more historically minded than any previous age."[11] Anxiety about our circumstances, and our identity, grows more acute the more we burrow into that sand dune of the written past. The nature of the primitive world is at the center of our modern anxiety about essence, appearance, and change because history cannot resolve for us the problem of change, which was mythically assured for many thousands of years as a form of renewal. Since we humans are not now what we once were—bacteria or quadruped mammals or apish hominids—other forms of life are irrelevant. The truth of history is that the more we know the stranger our lives become.

In the popular imagination our life in nature (everything outside this historical past) is in doubt, a shadowy and dangerous jungle from which we have escaped. In our search for ourselves, history narrows that identity to portraits, ideology, the adventure of power, and abstractions to which nations commit human purpose, to what feminists call "his-story." Carlos Fuentes writes: "Before, time was not our own, it was providence's own sphere of influence; we insisted on making it ours just so we could say that history is the work of man. . . . If such is the case we must make ourselves responsible for time, for the past and the future, because there is no longer any providence. . . . We must sustain the past, invent the future."[12]

History, like a biased science, verifies rather than demonstrates. Whether its narrative is interesting or horrible, its events are irretrievable as personal experience. It deals with an arc of time and measured location. Its creative principle is external rather than intrinsic to the world. Deity is distant, unknowable, and arbitrary. The historical past is the equivalent of a distant place in a cosmos whose first law is that you cannot be two things, in two places, or in two times, at once. It contradicts the fabulous tales, called "oral tradition," about an endless return. Having shaken off the garment of myth and put on the robes of dry history, we gain the detachment and skepticism that define the Western personality and civilization.[13]

✦

HISTORY REJECTS THE AMBIGUITIES of overlapping identity, space, and time and creates its own dilemmas of fragmentation and alienation—alienation from the domains of nonhuman life, primitive ancestors, tribal peoples, and the landscape itself. Living within this historic tradition, we find the meaning of life eluding us in certain significant ways.

Lacking a sense of the spiritual presence of plants and animals and of nonliving matter, we do not feel our ancestors watching or their lives pressing on our own as did prehistoric peoples. N. K. Sandars, an expert in prehistoric art, tells us that animals, as depicted in sculptures and cave art and reliefs, are never neutral. They carry meaning as "a profane source of food" but are also "sometimes a supernatural being, or even a god.[14] Historical consciousness gradually weeded out animal metaphors, organic continuities, and especially the perception of nonhuman spirits of the earth.

A repeated question of our time is, "How do we become native to this place?" History cannot answer this question, for history itself is the great de-nativizing process, the great deracinator. Historical time is invested in change, novelty, and escape from the renewing stability and continuity of the great natural cycles that ground us to place and the greater community of life on earth. As Norman O. Brown writes: "Man, the discontented animal, unconsciously seeking the life proper to his species, is man in history: repression and the repetition-compulsion generate historical time. Repression transforms the timeless instinctual compulsion to repeat into the forward-moving dialectic of neurosis which is history."[15]

In this new "Space Age" we are antigeographical. Place no longer exists as the womb of our childhood and the setting of myth. The economic unity of humankind, the multinational corporation, and the technology of travel and communication join us to all parts of the earth yet leave us homeless. Being largely placeless, the "world religions" belong to history, where "going native" is a misanthrope's hopeless escape or a "romantic nostalgia." Like thankless children, failing to acknowledge our connection to prehistory, we can live only in history, repressing our deep past as though it were an elemental irrelevance.

What was once the slow movement through habitats and terrains, enriched by narratives and the ongoing reciprocity with true residents, has been reduced to what Michael Sorkin calls "evocations of travel . . . places that refer to someplace else . . . the urbanism of universal equivalence," and electronic simulacra. These false landscapes, such as Disneyland, instead of containing secret reflections of our individual maturity, yield immature adults whose mythology is Mickey Mouse. Nature becomes a stage where the regimes and tales of power are enacted. To conventional history, technocracy adds the planetary imperialism of franchise business and the wasted landscapes of industrial and nationalistic enterprise, recreation as sheer kinesthetic motion, and the vacuity of the escape industries—as Sorkin puts it, the "celebration of the existing order of things in the guise of escaping from it."[16]

"Esthetic distancing," a distilled and rarefied concept of art passed on in Western culture from the Greeks, has become in our times an obsession with abstractions. Gallery art, stage drama, concert music—all so profoundly admired—are abstractions based on a logic of form. Virtuosity has become identified with celebrity and artistic excellence. Participatory arts that were once part of everyday life have become performance with the majority of humans in a spectator role.

Music is fundamental to our wholeness, our sense of primordial multiplicity. But observe what has happened to it in our time. The exaggerated solemnity of music in temples, churches, and mosques is a measure of the loss of joy and of organic sound basic to hundreds of indigenous religions marked by "mythic" imagination, the use of the skin-and-wood drum and group improvisation. Making music is often completely absent in the lives of our children.

Esthetic distancing also made possible the landscape arts and connoisseurship and commercialization as scenery painting, tourism, and recreation. To the credit of the Greeks, they resisted converting the landscape into scenery and wilderness into an aesthetic experience. In the sixteenth century pictorial space was invented by coupling mathematical perspective to painting. Nature itself became a kind of medium for highbrow entertainment, the pleasure derived would be ruled by artistic theory. The observer moved through life as though in a gallery.

Along with pictorial space and euclidean time, the phonetic alphabet was an inadvertent cause of estrangement. It made words an ultimate reality and the exposition of time linear—beginning with the bookkeeping mentality of the ancient Near East. The Mesopotamian desert-edge agrarians, and their "Persian" heirs of the mind, divided the world into material creation and infinite spirit that would shape the philosophy of the civilized world. Much of what we call "Western" has its roots in Hebrew supernaturalism and Greek hubris, behind which lurks the hieroglyphs of barter.

Elsewhere I have tried to describe history as a crazy idea, fostered not as an intellectual concept so much as the socially sanctioned mutilation of early childhood experience by blocking what Erik Erikson called "epigenesis," the complicit outcome of inheritance and environment.[17] Through education, history corrupts the intrinsic expectation of prehistory. Young children show natural tendencies that have always been part of the mythic mind as they personalize experience and show an intense interest in the natural world, especially in animals. They cling tenaciously to the proclivities that we try to educate out of them, the natural impulses that are the fundamental source of their creativity. Edith Cobb wisely said of childhood that its "purpose is to discover a world the way the world was made."[18] Children are in tune with that world. We personally experience childhood as a yearning, an intuition of the self, as other selves and other beings, a shadow of plant and animal kindred, vestiges of community that haunt us, and a need for exemplary events as they occur in myth rather than in history.

Most people most of the time in the history of civilization have lived under tyrants and demagogues, cued to despair and hopelessness. Today we are subject to progress, centralized power, entertainment, growth mania, and technophilia that produce their own variety of "quiet desperation." This desperation arises not only from lack of attachment to place but also from lack of kinship with the larger community of all life on earth. History is not a neutral documentation of things that happened but an active, psychological force that separates humankind from the rest of nature because of its disregard for the deep connections to the past. It is a kind of intellectual cannibalism which creates from those different from us a target group that becomes the enemy, upon whom we project our unacknowledged fears and insecurities.

History's judgment of the primitive world is a litany of excuses why we cannot go back: Time is irreversible. There are too many people on earth. Commitment to technology and its social and economic imperatives cannot be overturned. We cannot abdicate our hard-won ethical and moral achievements. Why surrender to a less interesting, cruder, or more toilsome life? History declares independence from origins and from "nature," which is outside the human domain except as materials and the subject of science. Politics that considers our dependence on the health of Planet Earth a moral imperative gives in to the rapacity of self-indulgence and egomania. In Philip Slater's words: "History . . . is overwhelmingly, even today, a narration of the vicissitudes of, relationships among, and disturbances created by those inflamed with a passion for wealth, power, and fame."[19]

OUR WESTERN EXPLORATIONS on this continent—our attitudes and consciousness as depicted in our conquest of the land and its indigenous people and our art—have been influenced by an unacknowledged aspiration lodged deep in our psyches and passed on to us from our European forebears: the search for a lost paradise. This longing for a perfect world may be the greatest motivator of our insatiable desire for the "good life." One wonders whether it is even possible for us to write about the past without a vagrant nostalgia for which perfect world that beckons to us but, so far as we can tell, never existed. History does not resolve our confusion but further misleads us with its mix of dreams and visions, infantile mnemonics, Golden Ages, Christian paradises, escapism, ethnographic misinformation, and fundamentalist attempts to make of it a mythology.

Christopher Lasch gets to the heart of our confusion: the distinctive conception of history is associated with "the promise of universal abundance." Only in the twentieth century did we make "the belated discovery that the earth's ecology will no longer sustain an indefinite expansion of productive forces." The notion that recorded history is an unfolding of human capacities, that we are heirs "to all of the achievements of the past," runs "counter to common sense—that is, to the experience of loss and defeat that makes up so much of the texture of daily life."[20]

Schneidau has told us that myths "do for the group some of the things that dreams do for individuals." Myth and the unconscious are the sources through which we access our numinous past and ease ourselves out of fears and contradictions into "mental patterns that can be dealt with."[21] By discrediting the importance of myth, the ideology of history has corrupted basic human thought processes that have been enriched by myth since we became human. There has been an educated genuflection before the idea of myth since Carl Jung and then Mircea Eliade and Joseph Campbell attempted to demonstrate that myth is a narrative expression of universal internal archetypes. But, in general, myth has come into ill use and has been depicted as stories that are false, beyond comprehension, or unbelievable. "It is only a myth," we say of stories too fanciful for reality. As a result of this general disrepute, it is difficult for many to credit "factual" history as the new myth of time and progress.

Jean-Paul Sartre argued that the dialectic way of approaching conflicting points of view by thoughtfully resolving contradictions is precisely what distinguishes civilization from the savage world. Sartre's mistake, says Claude Lévi-Strauss, makes him no more sophisticated than a Melanesian native who insists that the only stories that truly explain the world are his own. Lévi-Strauss argues that history is a myth because there is no possibility of recapitulating everything that happened, so history concocts its own story. Moreover, history is not a true sequence. It is fallacious to conceive history as a continuous development beginning with millennia and then going on to centuries, years, and days. These different time frames are separate domains, the larger units characterized by explanation, the smaller by information. As Lévi-Strauss points out, historical thought is analytical and concerned with continuity and "closing gaps and dissolving differences" to the point that it "transcends original discontinuity." In contrast "savage thought is analogical" and its main feature is "timelessness." Lévi-Strauss characterizes the source from which the "savage mind" draws its knowledge as a room with "mirrors fixed on opposite walls, which reflect each other. . . . A multitude of images forms simultaneously, none exactly like any other, so that no single one furnishes more than a partial knowledge . . . but the group is characterized by invariant properties expressing a truth."[22]

If not to the "historical consciousness" for the truest meaning of life on Planet Earth, then where are we to turn? Perhaps the prehistoric unconscious forms a better basis for the creation of a new history.

NOTES

1. Octavio Paz, *The Other Mexico: Critique of the Pyramid* (New York: Grove Press, 1972), p. 76.
2. Herbert N. Schneidau, *Sacred Discontent: The Bible and Western Tradition* (Baton Rouge: Louisiana State University Press, 1976), pp. 76–78.
3. Ibid.
4. John Cobb Jr. *The Structure of Christian Existence* (Philadelphia: Westminster, 1967), p. 82.
5. Schneidau, *Sacred Discontent,* p. 77.
6. Cyrus H. Gordon, *The Common Background of Greek and Hebrew Civilizations* (New York: Norton, 1965), p. 19.
7. Cobb, *The Structure of Existence,* pp. 52–59.
8. Ibid., pp. 73–75.
9. Ibid., pp. 124–125.
10. Editor's Note: As Schneidau has noted, the silence in the Bible about "the great dream of the mother goddess, which dominated the near East for many centuries," speaks volumes. See Schneidau, *Sacred Discontent,* p. 62.
11. Herbert J. Muller, *The Uses of the Past* (New York: Oxford University Press, 1952), p. 38.
12. Carlos Fuentes, *Christopher Unborn* (New York: Farrar, Straus & Giroux, 1989), p. 280.
13. Robert Hutchins, Preface to Mortimer J. Adler's Hundred Great Books Series, *The Great Ideas* (Chicago: Encyclopedia Britannica, 1952).
14. N. K. Sandars, *Prehistoric Art of Europe,* 2nd ed. (Middlesex: Penguin, 1985), p. 70.
15. Norman O. Brown, *Life against Death: The Psychoanalytical Meaning of History* (Middletown, Conn.: Wesleyan University Press, 1959), p. 93.
16. Michael Sorkin, "See You in Disneyland," in Michael Sorkin, ed., *Variations on a Theme Park* (New York: Hill & Wang, 1992), pp. 205–232.
17. "Epigenesis" is a term used by Erik H. Erikson *The Life Cycle Completed: A Review* [New York: Norton, 1985], pp. 26–27 that he borrowed from embryology to denote the phenomena linked with a creature's growth and development. See also my *Nature and Madness* (San Francisco: Sierra Club Books, 1982).

18. Edith Cobb, *The Ecology of Imagination in Childhood* (New York: Columbia University Press, 1977).

19. Philip Slater, *Earthwalk* (Garden City: Anchor, 1974), p. 156.

20. Christopher Lasch, *The True and Only Heaven* (New York: Norton, 1991), pp. 528–530.

21. Schneidau, *Sacred Discontent*, p. 7.

22. Claude Lévi-Strauss, *The Savage Mind* (Chicago: University of Chicago Press, 1966), pp. 260–263.

II

Getting a Genome

HUMAN EVOLUTION is a long, tangled tale that ties us inextricably to everything on this earth and, plausibly, to everything in the universe. In this day of Darwinian sensibility it is no more necessary to defend biological evolution than it is to defend the roundness of the earth. It seems evident that our genome—the sum of an individual's genetic material that constitutes the forty-six chromosomes in humans and controls heredity—is a product of millions of years of evolution.

We began as the species *Homo sapiens* in the Pleistocene about 500,000 years ago, but our genome is as old as life itself. Imagine the human genome, composed of chromosomes passed on to us, one-half from our mother and one-half from our father, as a precious heirloom made up of jewellike strings of genes, composed of DNA, nucleic acid combinations, that determine the way we look and function biologically and predetermine to some extent our potential. Because of the vast possibilities for our parents' chromosomes to divide and recombine, each of us, except for identical twins, is born with a different genome. The source of this genetic material has been passed on to us not only through our parents and generations of humans, but from archaic ancestors: primate, mammalian, reptilian, amphibian, ichthyian, and down to bacterial forebears of life on earth. The specific human part may be imagined as composed of diamondlike genes nestled in clusters of primate pearls, which in turn are distributed among a massive, gemlike heritage of still older ancestral markers. Recent genetic research showing commonalties of our genes with other species substantiates our innermost feelings that we all came from the same source.

We are not, however, what we always were. Genetic change does occur and can be extremely rapid in small, intensely selected populations—as in the remnants of a decimated group, with some island populations, or among domesticated plants and animals. But the evidence of genetic change in hominid paleontology is consistent with the slow rates of change occurring in wild populations, probably on the order of a few gene changes per 100,000 years. As a result, our Pleistocene specieshood owes little or nothing to evolution during the last 10,000 years, except perhaps for some local shifts in gene frequencies associated with resistance to epidemic disease, food allergies, or crowding, along with a widened diffusion of genes among races that were isolated earlier on.

The sweep and surge of modern evolutionary studies and the sallies and feints among anthropologists debating our human ancestry are to an onlooker like the crisscrossing tracks of a herd of restless wildebeests. Our archaic genealogy seems to have begun with the prosimians and their preoccupation with group life that is central to human identity. They were followed by the Old World arboreal simians, monkeys who divested themselves of ancestral prosimian dependence on the sense of smell but who elevated the social nexus to new levels.

When the primates came down from the trees, becoming in part or wholly terrestrial, as some 150 species have done, things happened in an interlocking fashion. Fossil bits and pieces begin to show a family of eighty-pound hominids, various species of what are now called *Australopithecus.* The big toe came in line with the other toes as the pelvis and legs made more dramatic changes than the shoulder girdle. There was a shift from quadrupedal to bipedal locomotion, more specialized use of feet and hands, and an increase in body size and accommodations in body shape to the upright position. Early bipedality emerged in complicity with the bones of our pelvis and feet. Sexual dimorphism, the differences between males and females, appeared. Social organization tightened up. Analysis of dentition reveals that what our ancestors ate is probably still best for us to eat and illuminates the reciprocity of tools and teeth and jaws, tongue, and pharynx that made possible the emergence of speech.

These major human modifications and adaptations are probably related to the emergence of human bipedality. As C. L. Rawlins puts it, "I'm a primate evolved for foraging the African savannah. My basics—legs, eyes, hands—are suited to light scavenging. My eyes are good at picking up quick movement, the flop of vultures from a lion kill or the scuttle of rabbits into brush. My hands are good for wrenching the joints of carcasses, prizing roots from the earth, plucking leaves and berries. Like my hands, my digestion is able to handle a wide variety of things."[1]

Robert J. Blumenschine and John A. Cavallo have suggested that among our early hominid ancestors, "scavenging may have been more common than hunting two million years ago at the boundary between the Pliocene and Pleistocene epochs."[2] Because of our tendency to "project current ways of life into the past," many anthropologists have failed to see the advantages of scavenging in our archaic past. But in terms of energy expended as compared with caloric intake, scavenging of dead animals makes sense. Mixed groups or individuals of these first hominids were probably expert at exploiting the immediate environment by scavenging dead carcasses, gathering all sorts of vegetation, insects, and larvae, and snaring or catching small game and fish. It follows that this sort of foraging activity would precede individual and group hunting of large mammals until strategies and know-how made it possible to procure large game without expending great quantities of energy.

A heritage of climbing ability may likewise have preadapted these hominids to stealing antelope kills stashed in trees by leopards or watching lions and jackals in the hunt while perched safely aloft. Vultures, in an extensive net of soaring individuals, watch each other, so that around birds descending on a carcass a centrifugal vortex is formed that may draw others from hundreds of miles away. A smart terrestrial scavenger and good runner, watching the vultures from the ground, might cover several miles in time to benefit.

Night restlessness, typical of terrestrial primates, may be a precaution against dangerous predators. It may also have been a way in early primates of recognizing the sounds of panicking prey or roaring carnivores at night—in order to remember the direction of kills that could be scavenged. This implies the perfection of mental maps and the ability to picture the known terrain. Finally, foraging in groups and carrying sharp tools may have furthered the transition from scavenging to large-animal hunting.[3]

As Blumenschine and Cavallo point out, foraging societies are egalitarian and one can envision the free overlapping and reversal of roles of female and male hunter/gatherers depending on circumstances. What group hunting brought was the development of sharing and cooperation and increased capacity to communicate among each other and to read the body language and other signals among dangerous carnivore competitors. A division of labor appears to have developed very early in our ancestry—it exists in most foraging cultures today—and must have advantaged survival. Children of both sexes would have had a basic grounding in scavenging strategies and an understanding of the distribution and appearance of plants and animals with the seasons. Likewise, they would have developed skills in locating carcasses before competitors, such as hyenas and vultures, or observing them and driving them off.

Standing upright opened the way for a more dexterous use of forearms, so these archaic forebears could not only stand up like chimps and bears, but could also run and carry things. If it was not for carrying babies in their arms, why would this capability emerge when a chimplike prototype did so little carrying? The human lack of hair to which babies (like little chimpanzees) might cling made that mode impossible, as did the added disadvantage of the jolts created by running upright.[4] Was uprightness also for carrying spears, escaping/pursuing in open country, seeing over tall grass, picking the seeds from high grass? The need to carry things to a central camp by these socially cohering food-sharers may have prompted woven bags long before any records indicate. Moreover, sunglare is an important limitation in bright, open-country savannas where humans first emerged. One wonders whether the heavy-browed ape skull was not preadapted to giving eyeshade to our ancestors who by that time had their hands full.

We come from a long line of primate omnivores. Just exactly when our teeth took distinctive shape and how this was related to scavenging and predation, the emergence of speech, and use of tools is part of the intellectual mix of modern anthropology. In our past history, primates and ourselves had some common ancestry, and we share much with those present on earth with us today. Among contemporary nonhuman primates, one or another species hunts, shares, cooperates, carries, keeps kinship ties, divides labor sexually, prohibits incest, makes tools (in both gathering and hunting contexts), shows linguistic capability, or has a long memory. All together these are the greatest story ever told.[5] We also share habits of eating that include consuming "flesh" in various forms. Anthropologist Robert Harding has shown that 69 percent of all primate species deliberately eat invertebrate or vertebrate foods. "Primates," he says, "can only be described as omnivorous; they are definitely not vegetarian animals."[6] Shirley C. Strum, in her studies of baboons, has observed them cooperating to hunt other mammals, sharing the kill, and carrying the kill to eating sites.[7] As recorded by anthropologist Geza Teleki, chimpanzees while hunting utilize "cooperative production," all spacial dimensions of their habitat, food sharing, and division of labor.[8] Hot food, the "warm meal" of which the raw, freshly killed animal is the prototype, is the sine qua non of the palate. The main difference between human hunters and today's nonhuman primate predators is that the latter do not hunt prey larger than themselves.

WE CAN TRACE OUR PROSIMIAN ORIGINS, anthropoid kinship, the shared skeletal, dental, and neurological features of our family, the Pliocene hominids, from which our own branch made its departure more than five million years ago. About two million years ago our ancestors emerged from their Australopithecene preamble as the genus *Homo*, bipedal, with a chimpanzee-sized brain, poised on the brink of an ecological adventure unknown to the other primates.[9] Out of that past emerged the great variations and races found in humans on earth today.

A useful way to think of this sequence of events is not in terms of taxonomy but of significant shifts in mobility and diet. Many who write on human evolution emphasize rapid changes rather than the slow evolutionary grind implied by "mutations." The mutated forms of genes do not usually swing into instant action to produce physical realization or "phenotypes," forms with visibly different characteristics. They remain in the gene pool sometimes for thousands of years, being eliminated on a regular basis, unless they find the right environmental circumstances—at which point they become visibly expressed in the creature's physical characteristics or behavior.

The biology of a basic human genome does not contradict variation. Nor does it imply that one race is better than another except in the context of specific environmental challenges. Psychological differences may occur between populations and races, since cognitive abilities are related to specific tasks and may vary due to selective pressure in different environments. The key to the diversity of human behavior may yet have its origin in the primate past—in a peculiar bimodality of the genes that may open us to possessive, competitive, cultural expression or to the more cooperative sharing cultures typical of Pleistocene peoples.

Both aggression and cooperation may be intrinsic and available to various human economies. In his book *Social Fabrics of the Mind*, primatologist Michael R. A. Chance suggests that two basic tracks were open to different primates in their social relations. Human evolution, being heir to a wide range of different species of ancestral primates, bears vestiges of both realms, making the human personality subject to alternative possibilities. These two "mental modes," combining brain structure and social relations, he calls the Agonic and the Hedonic.[10]

The Agonic personality is typified by the rhesus monkey. Its attention constantly flows toward the higher rank, making it "centrist" in its orientation. Unprovoked aggression, threat, and reconciliation give society its pulse. Low rankers get back into the group by "reverted escape" and submission. High tension characterizes the group. Sex becomes symbolic of power. The Hedonic personality is more like that of the chimpanzee with its appeasement, reassurance, and mutual dependence. The normal arousal level is low, there is little aggression. Threat is subject to reconciliation and reunion. The group has a strong general sense of unity, even though it may appear in disarray to an outsider.

This bimodality, says Chance, is "deep-seated in our nature." Its outcome depends on the social system that cues us and to which we apply its logic. If our "way of life" is efficaciously described in terms of its diverse relationships, then it can meld these two opposing personal and social modes in gradations, even distinctive ratios. Hunting/gathering, says Chance, is based on an equality principle expressed in fluid, reciprocal, social relationships and role integration that values teamwork. But its egalitarian style can regress into a rank-dependent, self-defensive arousal focused on self-security if it is unduly stressed, just as it does in chimpanzees.

The world of both chimpanzees and human foragers is typically safe and sufficient. Groups are open and followership is voluntary without fixed leadership, small groups dissolving and reforming differently. Individuals are normally calm and charismatic. All returns are immediate—as distinct from delayed return systems with storage in which cultivation of the soil and rights over assets such as boats, traps, and structures are protracted. The Hedonic system of mutual dependence, confidence, trust, and good-natured mutual assurance is not, says Chance, just a social creation of humans but a basic biological mode among certain primate kin, just as the competitive, aggressive mode is also in our genes from yet other relatives.

THE MOST EXTRAORDINARY FEATURE of human evolution is ontogeny—the specialized and scheduled development of physical and psychological traits that appear, disappear, or stagnate during the life cycle of the individual. *Onto-geny* literally means "the genesis of being." Of all the biological characteristics of humans originating over the millions of years of our later primate ancestry, and disastrously ignored in our perception of ourselves, ontogeny sets the timetable of the whole individual life.

Mice and other nonhuman animals, whose life cycles were first studied by biologists, do not change conspicuously after sexual maturity, nor do they live long. Most species of animals produce large numbers of young that develop rapidly and, typically, die before becoming adult or have just a brief existence as an adult. As a result of the early studies on animals, ontogeny is usually narrowly defined merely as the period from birth to physical maturity. But this is a misconception in terms of human development. Psychological changes continue in humans long after we have matured physically.

Some species, including our own, give more time and energy to these sequences of biological imperatives—for example, we humans devote a great deal of care to slowly developing our young whereas rodents invest in numerous progeny and thus accommodate high mortality rates. Human life stages cover seventy or so years, during which span traditional societies recognize a rich sequence of passages and roles. Human longevity is often misunderstood simply as more time to play, to grow, to learn, as though we had an add-on gift for getting more time out of life. Social support during these life stages is the yeast of maturity, the mentality of growing up.[11] Perhaps much of the violence, identity crises, and family disruptions in our time begin because we do not attend to the genetic "expectation" of such changes in our lives as the individual faces developmental challenges.

In childhood there are three neurobiological stages of mental representation: enactive, iconic, and symbolic. The first stage, the enactive, is basically mammalian and is largely sensory—body movements that trace first their mother or caregiver's body, its smell and feel, and then the larger environment around them, the movement through space, and the imprinting of place and its components (weather, water, rocks, soil, plants, animals, people). In this beginning stage, children explore their way through their environment much as rats in a maze or animals in their natural habitat. Touch, smell, and hearing are especially important in this most fundamental orientation. Iconic representation takes place in a series of signs and images of increasing complexity. The figures of humans drawn first by young children—arms and legs with fingers attached to a head—that progress as the children grow older to more realistic representations of the human body illustrate the progressive development of this sign world. Speech is part of this mode of understanding because the icon represents some part of the environment. The child early on knows what a "bow wow" represents and does not confuse it with a "moo cow." And later the child can distinguish between dogs (and people) by ascribing proper names. The third form of representation is symbolic—a way of referring in which the symbol may lose any similarity to what it stands for (an "overdetermined" metaphor) and its "meaning," therefore, must be learned and taught. These three stages of representation follow a hereditary ontogenetic agenda.

Psychologically we are ontogenetic as well: the personality and temperament follow characteristic patterns and needs laid down over the Pleistocene. While conventional psychology and child development recognize this ontogenetic imperative, they have seldom asked where it comes from, why it is there, or how it was adaptive in our evolution.

More is known in child care about the failure of the enactive phase because it is so much a part of mother/infant relations and the psychopathologies that follow its failure. The absence of a functional iconic basis in nature impairs our sense of the diversity of life or the implications of terrain, earth, and its life. Ontogeny does not imply that we redress our errors by feeling good about nature instead of fearing or wanting to control it. As a highly specialized life form, we are genetically endowed to "expect" fulfillment of the genome's childhood schedule of needs and abilities, to which society is tutor and guide with its tests, informal daily life, and formal ceremonies that erupt and fall in time like the successive molts of feathers on the body of a bird. Our extended human ontogeny, with its natural demarcations in stages and phases, is governed by neoteny (a "state of newness")—a retardation of certain parts of the maturing process. Neoteny preprograms life stages, so that our becoming is a lifelong process.

We, among all creatures, are in some ways the most free. Yet, even though blessed with wider choices than the other animals, we are not truly free to be immature, or for culture to neglect to mitigate our immaturity. That modern psychology has taken the wrong track is reflected in the popular narcissism of the self and the study of the personality as though adolescent self-absorption were normal in the context of the hubris and hedonism of our affluent society. Modern psychology, including "eco-psychology" and "environmental psychology," tends to portray the self in terms of individual choices about beliefs, possessions, and affiliations rather than defining the self in terms of harmonious relations to others—including other species—and in terms of the ecological health of the planet.

Ontogeny includes a synchrony of brain and neuromuscular development that corresponds to the wants and needs of the individual. Culture—a heritage of skills, attitudes, traditions, language, and arts—

evolved out of a biological potential embedded in this ontogenetic agenda. Cultural responses to our inherent development as individuals have content. They have been worked and reworked so long that there is empirical wisdom in the social and cultural mentoring of the individual. The "extended childhood" and the characteristics of the adult that carry youthful traits into later life are therefore at the heart of human biology and evolution. The agenda is a given; the support depends on a social readiness to nurture, itself a product of successful ontogeny of an older generation. To the often asked question, "Why don't you grow up?" perhaps the answer should be, "Because I need a bit more time and understanding."

✦

I PROPOSE that our ontogenetic agenda has been carried in our genome from Pleistocene times when our species made its debut. Furthermore we have inherited from our primal ancestors an orientation to the world, a way of perceiving our place in the scheme of things. Let us go back for a moment to a long view in order to retrace those first steps toward our present humanity.

About two million years ago, at the border of the Pliocene and Pleistocene epochs, our first ancestors, *Homo habilis*, moved out toward the forest edge. Following corridors of riparian woodlands, where, being partially arboreal, they could in time of danger seek the sanctuary of trees, they could also make excursions into open country. In such a habitat they would have access to grass seeds, ground-nesting birds, certain reptiles, young mammals, and carcasses from big cat kills. Eating grass seeds may have stimulated the upright stance—freeing the hands—which would have helped not only gathering but general lookout in tall grass. The number of large, dead bodies available to these human scavengers would have increased in the savanna ecosystem not only because of the larger number of animals and their predators but also due to natural death in the dry seasons. Big dead bodies, moreover, last longer than small ones. Meat is a welcome food among higher anthropoids, and the opportunities may have sharpened their attention to the behavior of carnivores and strategies for finding and using flesh of carcasses. Windfalls of meat from large animals were advantageous mainly if some of it could be carried away—served not only by bipedality but also by ahead-thinking and sharing.

The oldest known tools coincided with the earliest members of the genus *Homo* when they probably began to cooperate in foraging and food sharing as they scavenged in the riparian woodlands or open areas. These were percussion pieces for the extrication of roots from the earth and meats from nuts and for smashing bones. Early on, before cutting tools, human scavenging would have depended on the dead whose body cavities had already been opened or whose meat had been stripped. Percussion stones were used for breaking the larger bones to get the marrow and for opening the cranium to extract the brains. The sharp edges of fractured bones may have served for cutting, preceding the use of shaped stone for defleshing dead bodies. Competition with other scavengers and the ability to drive carnivores from their kills would have also facilitated the evolution of weapons. There was, however, not a single tool of the Pleistocene apparently made for war.[12]

To cut more precise pieces, sharp flint edges, choppers, axes, and stone flakes appeared. Humans began looking inside animals, opening bodies, noticing that parts of different animals corresponded, the parts themselves becoming "species" with their own taxonomy. We made the marvelous discovery that inside we and the others were even more obviously kin than indicated by our exteriors. But in scavenging dead bodies, we never abandoned a general subsistence that kept our omnivore bodies healthy and turned our attention to the whole landscape. Gathering and scavenging, as noted earlier, are not distinct activities separated from hunting, nor do they require less acumen. Scavenging large animals requires many of the same skills needed in hunting.

Prehuman foragers, even in their earliest centuries at the edge of the forest, were never so stupid as to simply ramble about, blindly following chance probabilities of encounter, rather than exerting the kind of superior intention that is obvious in all primates. Avoiding and outsmarting predators, sensing where fruits are ripening and roots are plentiful, distinguishing poisonous herbs and fungi, recognizing plants with healing properties, being wary of signs of kills for possible scavenging, thinking ahead in terms of the need for tools for digging or breaking open bones—all this necessitated planning and fore-thinking. Developing a keen awareness of the environment and its potential for food sources, as well as the conceptualization of mental maps required higher mental skills than just reading signs. It also required passing some of the information on to progeny.

Hunters frequently stop to pick and pluck—to "gather"—eggs, turtles, frogs, and insects that may be eaten on the spot; a gathering group not only digs roots and picks berries but also kills small game and scavenges. While the origins of the intelligent human hunter/gatherer go far back into the social structures of ancestral lemuroids and the vocality and vision of arboreal simians, the breakout comes with savanna omnivory: now intelligence was fostered among bipedal scavengers carrying tools with cutting edges and finally cooperating to kill large game and share food.

Some crucial social and intellectual mileposts had to be passed in order for our ancestors to hunt cooperatively and share large animals. Our hunting began two million years ago with the 500-cubic-centimeter brain that reached 1,500 cubic centimeters about fifty thousand years ago—and with this increase in brain size came a concomitant ability to conceptualize. Very early in the story, for example, the recognition of other species at a distance would have simply extended a preexisting ability of all large vertebrates in the circumstances of savanna life. Soon after would come a quickness for attaching sounds or smells to those same species, even when they were not visible. And, like wolves, the human predators would have expanded these recognitions to subgroups within the species in order to know which antelopes or zebras were old, sick, pregnant, and very young and which would defend themselves dangerously or flee with ease. The success of carnivores—including the human as carnivore—is always marginal. Playing the odds is essential.

As our hominid ancestors increasingly moved into open country, often in sight of prey for hours at a time, it was possible to recognize a kind of daily round of other species, if one had the memory for it, to know when the prey slept, grazed, watered, or changed locations for special feeding, courtship, or bearing young. With forethought the hunter could be present in those places before the prey or ambush them along the way. Ambushing suggests the use of cover, and it is not difficult to guess that our forebears watched lions and cheetahs using not only the vegetative cover but rock and terrain to stalk, and learned to imitate them. At some point the idea of such observational learning must itself have become conscious, so that every species of animal became a potential teacher.

Nor would it have been only the predators who were seen as models: the cunning of prey must also have been the object of our ancestors' inquiry and admiration. At a weight of 70 pounds millions of years ago or 150 pounds thousands of years ago, they would have continued to be the object of the hunt themselves. Species of canids, cats, and hyenas now extinct as well as those with us today probably relished all kinds of primate flesh. We would not have endured as smart hunters if we were dumb quarry.

The canids and lions would surely have been models of cooperation in the hunt as well as in the division of labor. The aptitude for working together and not spoiling it at the end in conflict would have been, then, more than a mere discovery. It would have required new social skills and understanding. Individual personality would surely have been a large part of this awareness. Having learned from the animals and the nonliving surround, our primal forebears emerged from the Pleistocene wary, able to discern advantage in chance encounters as well as skilled in planning ahead, keenly sensitive to the environment and its signs, communicative, cooperative, and sharing.

THE KIND OF INTELLIGENCE and cunning needed by our primal ancestors to develop and survive as they did during the Pleistocene has been overlooked. Or, worse, it has often been translated into a condescending attitude toward modern aboriginal people who are seen as "savages." The cruelest form of modern criticism of primal peoples depicts them as stingy and greedy as anybody else, implying that to be human is to be selfish.

The most strident of these theories projects overkill onto the aboriginal inhabitants of the world by claiming that, being basically avid, they were responsible for the extinction of many large animals at the end of the

Pleistocene. Invading hunters from Asia, the argument runs, exterminated the giant sloths, mammoths, and horses. Their relentless pursuit of hapless and trusting animals who had never seen humans presents a portrait of grisly slaughter indeed. We are encouraged to picture cliffs where men drove bison or horses to their deaths—a kind of epigram for the whole sordid episode of the hunters' blood lust.

Studies of hunting/gathering peoples show that to hunt big mammals exclusively is bad strategy: generalized subsistence is more efficient and reliable; and indiscriminate hunting is inefficient and goes against long-term survival. Indeed, the proposed high predation of megafauna, such as elephants, among prehistoric peoples is extremely naive if one considers the time and labor necessary for hunting large animals. Archaeologist Raymond E. Chaplin says, "Prehistoric man is unlikely to have created any strong imbalance or brought many species near extinction."[13] Except in the Arctic—where the animal fats are polyunsaturated and sea ice and sea strand hunting is efficient—exclusive big-game hunting is unwise and inefficient. Most of the documented extinctions brought about by primitive humans are associated with islands or with agriculture. Donald Grayson explains that during the last few thousand years of the North American Pleistocene, as many as thirty-two genera of mammals and ten genera of birds became extinct. He argues that this episode of extinction is too narrow and the variation of extinctions too wide (ranging from blackbirds to mammoths) "to be accounted for by . . . human predation."[14] For these reasons the Mosimann-Martin model that presents the hypothesis of prehistoric overkill by humans is not convincing.

Of the known Pleistocene extinctions only 9 percent of the thirty-two extinct genera occurred during the late Pleistocene with the human advance into North America. Some 50 percent of the extinctions took place during the Gunz glaciation and 25 percent during the Riss-Wurm glaciation. Just prior to the human arrival in North America, about twelve species of megafauna vanished. Among them were a huge carnivorous bear, a gigantic lion, two genera of saber-toothed tigers, the jaguar, a cheetah, and the dire wolf. There is virtually no evidence, such as stone implements, of confrontation between humans and most of the extinct animals.

Peoples entered North America from Siberia along with the grizzly bear, moose, caribou, wolverine, wapiti, and bison, all of which survived. Their Siberian hunter descendants are known, like most such peoples, to limit their kill to the little they can store and carry. Coming from Asia, the migrants had behind them a long coexistence with the megafauna of Siberia. Conservation, not overhunting, was practiced among aboriginals—if not for ethical reasons then surely for practical purposes. In terms of caloric intake as well as energy conservation, it was advantageous to hunt species that were plentiful and to utilize multiple sources of food.[15] Overkilling was regarded with repugnance because it led to competition and territoriality between tribes and for these reasons was virtually unknown among hunter/gatherers.[16]

Little evidence exists, then, that humans were responsible for the extinction that took place at the end of the Pleistocene. In northern Asia the extinction of the mastodon and mammoth was associated with the diminished tundra; there is virtually no evidence of associations with bone accumulations. Valerius Geist proposes that humans are unlikely to have killed off the "densely-packed fauna of specialists" that became extinct on the North American continent. In fact, he suggests that the presence of these animals may have delayed human migration: two species of the sabre-toothed tiger, a huge lion, the dire wolf, and two species of the big *Arctodus* bear may have deflected the human passage down the west coast of North America.[17] Human overkill envisions a "front" of advancing human invaders, but no such pattern of migration existed; the first human inhabitants followed vegetational and geological corridors and coasts in streamlike, not wavelike, movements. In Eurasia major game animals such as the reindeer, red deer, auroch, and horse did not become extinct at that time, while less desirable game, such as the cave bear, rhino, and mammoth, did.

Perhaps the idea that our hunter ancestors extinguished many animals appeals to our Judeo-Christian apocalyptic imagery and our chronic modern guilt from having ravished a continent. It attracts a misplaced sentiment for preserving and protecting individual wild animals. Perhaps, above all, it resurrects old fictions about primal, "barbarian" peoples as rabid animals.

✦

THE ELEGANT REFINEMENTS of our species, as inherited in our DNA, are difficult to see because the genome is an unseen actor behind our daily behavior. Like a soft-spoken elder, its unique role is to call upon human society and imagination to invent its exact expressions. Each human group responds differently to the needs of the life cycle, the modes of foraging, the nature of the divinities, and the play of tropes and art.

Human societies vary greatly in their structure, but the differences, however crucial they seem to us, are variations on the species theme—whose *human* traits are Paleolithic. The health of a society is a measure of its freedom from stress, individual suffering, psychopathology, tyranny, and ecological dysfunction as a result of straying from that basic ancestral form. The greater the degree to which a person or society conforms to our Paleolithic progenitors and their environmental context the healthier she, he, they, and it will be.

NOTES

1. C. L. Rawlins, *Sky's Witness* (New York: Holt, 1993), p. 236.
2. Robert J. Blumenschine and John A. Cavallo, "Scavenging and Human Evolution," *Scientific American*, October 1992, pp. 90–91.
3. Ibid., p. 96.
4. The question of reduced hair in our species continues to be argued. The idea that it was for controlling body temperature does not square with the many hairy mammals who do not necessarily overheat. I find the logic that its purpose was erotic, part of the bonding system between male and female, to be more convincing.
5. Five thousand references are given in the library's section on evolution and anthropology. See, for example, Charles R. Peters, "Toward an Ecological Model of African Plio-Pleistocene Hominid Adaptations," *American Anthropologist* 81(2) (1979): 261–278, or John Gowdy, "The Bioethics of Hunting and Gathering Societies," *Review of Social Economy* 50(2) (Summer 1992): 130.
6. Robert S. O. Harding, "An Order of Omnivores: Nonhuman Primate Diets in the Wild," in Robert S. Harding and Geza Teleki, eds., *Omnivorous Primates* (New York: Columbia University Press, 1981), pp. 199–200.
7. Shirley C. Strum, "Processes and Products of Change: Baboon Predatory Behavior at Gigil, Kenya," in Harding and Teleki, eds., *Omnivorous Primates*, p. 263.
8. Geza Teleki, "The Omnivorous Diet and Eclectic Feeding Habits of

Chimpanzees in Gambe National Park, Tanzania," in Harding and Teleki, eds., *Omnivorous Primates*, p. 340.

9. Peters, "Toward an Ecological Model," p. 261.

10. Michael R. A. Chance, "Introduction," in Michael R. A. Chance, ed., *Social Fabrics of the Mind* (London: Erlbaum, 1985), pp. 3–9.

11. Erik H. Erikson, *The Life Cycle Completed* (New York: Norton, 1985), p. 31.

12. Julia Rutkowska, "Does the Phylogeny of Conceptual Development Increase Our Understanding of Concepts or of Development?" in George Butterworth et al., eds., *Evolution and the Development of Psychology* (New York: St. Martin's Press, 1985), pp. 115–129.

13. Raymond E. Chaplin, "The Use of Non-Morphological Criteria in the Study of Animal Domestication from Bones Found on Archaeological Sites," in Peter J. Ucko and G. W. Dimbleby, eds., *The Domestication and Exploitation of Plants and Animals* (Chicago: Aldine, 1969), p. 239.

14. Donald Grayson, "Pleistocene Avifauna and the Overkill Hypothesis," *Science*, 18 February 1977, p. 692. See also Karl W. Butzer, *Environment and Archaeology*, 2nd ed. (Chicago: Aldine-Atherton, 1971).

15. Michael A. Jochim, *Hunter Gatherer Subsistence and Settlement: A Predictive Model* (New York: Academic Press, 1976), pp. 16–17.

16. Renez R. Gadacz, "Montagnais Hunting Dynamics in Historicoecological Perspective," *Anthropologica* 17(2) (1975):149–168.

17. Valerius Geist, "Did Large Predators Keep Humans Out of North America?" in Juliet Clutton-Brock, ed., *The Walking Larder* (London: Unwin Hyman, 1990), pp. 282–294.

III
How We Once Lived

THE GENOME, our personal genetic constitution, is responsible for thousands of biochemical and physiological processes necessary for life. Beyond that, the human part of the genome has given us a brain for making choices and has also passed on certain expectations about the categories of those choices. Between the ages of nine and twenty months, for example our genome guides us to identify and name body parts and animals. As surely as it disconnects baby teeth in the sixth year, between the thirteenth and fifteenth year the genome calls for spiritual grounding—a spiritual experience that will connect the person to place and cultural heritage. Yet it does not tell us what language to speak, which body parts or which animals to name, or which tooth fairy to invoke. Bound by some intrinsic framework of social demands, we are set free to pursue the path to the fulfillment of these genetic requirements without specific rules of child care, of nurturance for each other, or of community design. This ambiguity between genetic requirements and freedom to express them is an astonishing aspect of human nature.

The human part of our genome came into existence along with social patterns and skills. And these were followed, over hundreds of thousands of years, by different human cultures, each unique and yet appropriate to the human niche. In a broad sense there was a Pleistocene way of life that encompassed the many human primal cultures, all of which were consistent in certain ways and are shared even today among recent hunter/gatherers. We are free to create culture as we wish, but the prototype to which the genome is accustomed is Pleistocene society. As a culture we may choose or invent any language or set of gods we like. But that we must make up a language and choose gods is what it means to be human.

Some cultures are more socially and ecologically attuned than others and produce institutions and perhaps even individuals who are less churlish, loutish, and brutish. Although cultural modes differ, one is, in itself, as "good" as any other. There is no inherent difference in the humanity of different peoples. We are a species. If the lives of some are better, it is because they live in a natural environment and a cultural system that are closer to meeting the "expectations" of the genes: the contract with evolution is being more generously fulfilled. Our world does not make us; nor do we make ourselves; we are the continuing creation of the interaction between our organic structure and the way we shape the world around us. It's possible to do it badly. It's also possible to do it well. We are an epigenetic phenomenon: our development is elaborated continuously during our entire lifetimes as it has been down through the ages.

✦

OUR BODIES ARE NICHE-FIXED, defined by the characteristic features of our ecology in the strict sense of the word—that is, the energy and symbiotic patterns and demographics of our genus, *Homo*, as they have existed for perhaps two million years. The genome has its demands, interpreted by the social structure of the primordial group. Above all, it links the individual's relationship to the social and ecological environment by the demands of a calendar bound to age. The highly specialized human brain and its delicately poised *locum tenens*, the mind, can perform extraordinary feats within this context . . . and an incredible variety of destructive derangement without it.

The "hard, irreducible stubborn core of biological urgency, and biological necessity, and biological *reason*," says Lionel Trilling, "reserves the right to judge the culture, and resist and revise it."[1] This core, I propose, guides ontogeny, the phases of each individual life, from conception to death. The expectations of our genetic endowment are not a gray, passive mass for registering whatever comes to mind, nor some compelling conglomeration of irresistible reflex arcs. Because of our evolutionary past and the extraordinary way life has shaped our mind and bodies, we are required by the genome to proceed along a path of roles, perceptions, performances, understandings, and needs, none of which is specifically detailed by the genome but must be presented by culture. Mentally and emotionally, children, juveniles, and adolescents move through a world that is structured around them following a time-layered sequence of mother and other caregivers, nature, and cosmos. Infants go from their own and their mother's body to exploring the body of the earth to the body of the cosmos. Our basic human intuition tells us that these bodies comprise a "matrix," that is, "mother." The significance of perceiving environments through a series of different but perpetually "motherly" matrices or contexts is that the world is prototypically organic, feminine, and maternal. The study of nature among primitives begins in childhood but is a lifelong preoccupation.

The most crucial human experience is childhood—its bonding, socializing, and exploration of the nonhuman world, its naming and identification. Speech emerges according to an intrinsic timetable. Language must be taught. But nature is the child's tangible basis upon which symbolic meanings will be posited. The naming and recognition of plants and animals of the home range is the primary function of speech in childhood and the basis for later metaphorical meaning. This is the first lesson: the basic mode of identification and understanding. Early speech is in terms of nouns and verbs. Talking—conversation—is not as important to small children as words for labels and categories, the skill of discriminating and naming without which no meaningful speech or higher cognition can take place. Naming at first involves body parts and then animals because anatomy is fundamental to all identity: body parts are the supreme objects for learning the skills of taxonomy. The ability to read the landscape or the environment, later in life, grows from establishing natural things as its anatomy, keys to the wholeness and well-being of the habitat.

Every change in a child's body and in its daily routine is anticipated and calls for shifting roles among its family, siblings, and others—changes not discovered anew each generation but anticipated by social experience and intuition. Timing is everything. Ontogeny involves not only physical traits, like tooth eruption or growth and change of hair color, but the personality as well. Neoteny, the specialty of our species, is mitigated by the readiness of human society to meet each stage of the child's development with an appropriate cultural response that is analogous to the optimal environmental conditions necessary for a butterfly to emerge from a chrysalis.

Neoteny, the immaturity factor, is intimately associated early in life with a topographic intuition—that is, place in consciousness as an aspect of one's own body and physiognomy. Early correlation in human subconsciousness between body and earth, as the child and then the youth explores and wanders through his or her home range, is basic to future visualizing of nonphysical reality and cosmological "places."

✦

THE SOUND WORLD as well as the sight world is extremely important to the development of a sense of place in children. Paul Sears, the renowned ecologist, is said to have told his children, "Never ignore a sound," whether it be the slight clicking of the first snowfall against leaves and grass or the sudden silence of frogs or birds. Not only separately but together, things have a voice. The Voice of life is made up of calls, drums, songs, musical instruments, moving wind and water; they tell us of the livingness of the world in a surprisingly coherent milieu. Vision discovers parts but sound links them. This process starts internally, like the rumble of an earthquake, becoming internal and external at once. Gary Snyder has called it "the primacy of together-hearing." Even percussive music and great intervals of silence are evidently conducive to our well-being. We have been surrounded now for millennia with domestic places that have become metaphors of a diminished self. Perhaps one way home is the path through music.

Those of us who continue to find coherence in music in a disorderly world may find it easier to open ourselves to music in new ways that would reconsider all sound as music. Sound therefore has a dual voice. The oldest myths of the origin of music tell of the mellifluous calls of grouse and certain pigeons, cackle of scrub turkeys, trumpeting of cranes, tempos kept by drums and bamboo jaws and rattles made of seedpods, mussel shells, gourds, or crayfish claws. The myths of human/bird transformations explain the categories of natural sound—weeping, song, poetry, whistling, talking, mimicry, noise. Nature is like a tuning fork: its space, time, and seasons are marked by an auditory pulse with its variations in echo and penetration, layers of the daily cycles of frog, bird, and insect calls. One sings in duets with the birds, cicadas, and waterfalls.

The Temiar, musicologist Marina Roseman tells us, are a rain-forest people of the Malay peninsula whose culture is filled with song and spirituality. "Instead of alienating flowers, trees, or cicadas as inherently different and distant," she says, "the Temiar stress an essential similarity." The Temiar "receive inspiration and constant regeneration from interactions with the essences of mountains, rivers, fruits, and creatures of the tropical rain forest. . . . Temiar culture is an exquisite translation of the natural environment into cultural terms. The jungle is a social space."[2] The emphasis on place is fundamental to their music. The forest is a reflection of social relationships mediated by song. "If we compare at the level of segmentary, nonhierarchical societies adapted to tropical forest environments," says Roseman, "two features become apparent. One concerns mutualistic responses to the rain-forest environment; the other, modes of political persuasion that are influential and cooperative rather than authoritarian and coercive."[3]

One can imagine what became of music in Western history.

We lost our informality in interaction and expression and song long ago, abetted along the way, as noted by author Dolores LaChapelle, by the mechanical clock promulgated to create and control a schedule of worship and work. In its thousand years the clock has become the great corporate destroyer of spontaneity. In her book on D. H. Lawrence, she refers to his *Plumed Serpent* regarding the replacing of bells in the church with drums: "In a few sentences Lawrence hints at the enormous changes implicit in moving from rigid clock time with metallic bells to drums timed to the natural rhythm of the day: dawn, first sun showing, sun highest in the sky and sunset." She goes on to say, "The bells call attention to the Christian Church standing there focusing all power onto itself; the drum connects humans with their circumambient universe and with nature's changing cycles." LaChapelle goes on to explain that "the drum is not a return to past ages; rather it is a remembrance of who we really are"—we lived to the syncopated beating in the womb, to the measure of the mother's heartbeat and that of her fetus which beats twice as fast. "The drum," she says, "has always been the center for sacred rituals in every culture in the world. . . . and the vehicle for ritual dancing as well."[4]

✦

CERTAIN BIOLOGICAL MARKERS in childhood indicate readiness in primitive cultures—for example, the loss of milk teeth is a signal that children are ready to accompany adults and help in foraging and caring for other children. The boy learns the rudiments of the hunter/tracker skills and the girl the intricacies of the digging stick and the subtleties of plant distribution, but no routine work is expected of either. Children listen with rapt attention to hunting stories that become for them a vast source of cultural information. They play at hunting and cooking, but are not pressed into the food quest. Nonetheless both sexes learn to identify hundreds of plants and animals.

Children at age six are typically anthropomorphic: they perceive other forms of animal life as motivated and feeling like themselves, which is the basis of kinship with the natural world. This feeling extends to plants as well. Trees are perhaps the most important plants in the lives of children. Because of our forest origins we have an affinity for trees, a tendency that is virtually compulsive in childhood and shared with most other primates. Gombe chimpanzees, for example, hunt game smaller than themselves, abetted by an open forest structure that hampers troop defense and enhances cooperative pursuit. The chimps climb trees for the view while hunting.[5]

It would be hard to overestimate the degree to which trees give internal shape to the space in which the child plays. They are on the one hand like great, protective, benign adults whose whispering and lightly percussive tremolo is like the humming of a kindly aunt or uncle. On the other hand trees structure space as though it were a labyrinthine underworld, where hiding is like survival itself.

Trees were made for climbing, a return to quadrupedal motion, touching a chord in our genetic memory of an arboreal safety. The rough texture of bark against the chest and arms, the smell reminiscent of a time so long ago that we still had whiskers, the gift of nests and fruit, the green galleries and corridors, the vestibular possibilities in being rocked by the wind or bouncing on a limb are part of my own childhood recollections that go deep. I remember, as a child, climbing a twenty-foot sapling until it bent gently and lowered me to the ground, crawling into the hollow trunks of big old sycamores or river birches, imagining the possibilities of something else being in there. Building tree houses like nests, like the spectacled bears in South America or chimpanzees in Africa making platforms, prompts delight at the thought of sleeping in treetops.

The dense forest has its gothic side and smell of danger too, perhaps as the visceral fear of an open-country vertebrate. The solitude, silence, dim light, and cool quiet of that great interior is profoundly calming. Tree climbing itself is very important. Colin Turnbull says: "A rich symbolism constantly reminds the Mbuti child of the supreme value of all, *ndura*, or 'forestness.'"[6] The forest's limitation of our movement is different from the sense of freedom that open country gives us, and its roots go much deeper into our past, as though the forest were part of our brain, a silent mnemonic reminder, especially for children who have not asked the question of meaning and continuity with the natural world, a reminiscence emerging later in life.

✦

ADOLESCENCE IS THE TIME in the life cycle when the human body develops to sexual maturity coupled with emotional changes that prepare the individual for adulthood in a community. Maturity is a relative matter. It depends on how well the individual has transited the passages of childhood and youth. Neoteny slows down development in some parts of the body and personality to the extent that certain immature traits remain throughout life.

Symbolic thinking comes with adolescence. Among primal peoples, plants and animals prepare the individual for the skills of metaphoric allusion to physical things in order to conceptualize abstractions. Totemism strictly speaking, is the social role of the individual by analogy to a natural series. Myths and spiritual and cosmological concepts are communicated by allusion to a familiar natural world.

Peer groups are unimportant in a band of twenty-four in which there may be seven or eight children of mixed ages. And older children caring for younger children may have important ramifications that are not yet widely understood. With primal people, there are no adolescent groups brought together for ceremonial initiation. (Adolescent in-groups and secret societies occur in competitive and warlike cultures, not among hunter/gatherers.) "Hanging out" together of age-stratified youths may be one of the most destructive characteristics of our present culture. Without a childhood that has grounded them in the natural world, often without adults anticipating and properly monitoring and celebrating their transition into adulthood and understanding their idealism and need for spiritual experiences, youth often find themselves alone in this modern world. In age-specific gangs they are "growing themselves up" the best way they know how, often in a milieu of violence and power rather than in spiritual communion.

Because of the modern sense of loneliness and lack of true community, many psychologists and counselors see "separateness" as a major problem of the individual. Anthropologists sometimes attribute the lack of anxiety about self-identity among tribal peoples to a lack of self-consciousness: they are thought to be sunk in a group identity like a school of fish. Others regard the struggle with individual identity as failure to "identify" with nature. Indeed, there is a movement among environmental philosophers to reenvisage the "self" to include plants, other animals, even the nonliving world. But maturity does not consist of the loss of one's body boundaries, a subjective prenatal universality. Normal development consists, rather, of sharpening the distinctions between the self and the other to clarify one's identity. The danger is that the self, constantly removed and made unlike the others, may become isolated. But the Pleistocene solution is the enhanced complexity of relationships. A healthy personal development proceeds through a corresponding process that emphasizes relationships to others, so that intensified separateness does not maroon but establishes the self as ever more unique and yet more fully bonded to nonselves by chains of interaction, kinship, dependence, cooperation, and compliance.[7]

In primitive communities a camp of thirty typically contains eight to ten hunters and the same number of gatherers. Prepubertal boys, like older men, help out and run snares. Prepubertal girls learn the skills by keeping at the sides of older women. Menarche is surprisingly delayed in girl foragers compared to modern females. Among the !Kung San of Africa it occurs at about seventeen years, in some industrial societies it comes as early as eleven. The reasons for the slower maturation may be normal exercise among the primitive people, reduced social intensity, reduced emphasis on sex in collective households in which children grow up in the presence of sexual life, or other factors related to diet or group expectations.

✦

MARRIAGE FOLLOWS as the individual shows herself or himself skilled in the day-to-day tasks of a foraging life. Marriage usually takes place at about age nineteen in primal societies, and the first child is born when the woman is about twenty. If the mother nurses the child for about three years she will probably have only three more children in her lifetime.[8]

Beyond the first twenty years are the stages of marriage, parenthood, young adulthood, midlife leadership, grandparenthood, kinfolk complexity, and elderhood. Each in some way is an appropriate aspect of the ontogenetic self, more or less fully realized and more or less nurtured and sustained by social skills. Meat is shared according to long-standing custom. The old are sustained and respected. The cycle of the seasons predicates the foods to be sought and the weather to be expected. Because hunter/gatherers move from camp to camp and build only tiny huts of sticks, their impact on the earth is almost zero. The various crises in modern society—schooling, adolescent gangs, divorce or enmity, midlife changes, the "care of the elderly"—may be only pathological expressions of normal ontogeny when the culture is so stressed and bent that it cannot guide the new person emerging at each stage.

Just as our bodies respond appropriately to a great many subtle stimuli during a twenty-four-hour day/night, we have different needs, anxieties, and social skills that emerge with age. What complicates this in our own time is the deformation of the life cycle, so that students of ontogeny face the difficult task of working backward from the stressed examples of children and adults and medically sustained old age.

✦

OLD AGE IS LESS WELL ONTOGENIZED, but it too has its nodes. Among the older generation of primal peoples, becoming a grandparent actually initiates a new level of child care, freeing the young parents for more strenuous duties that older people cannot do. It has long been speculated that our species has "postreproductives" because the old are better memory banks, keepers of the lore and genealogy, healers, accumulators of useful social lessons (especially childrearing and the resolution of disagreement) and are better suited to roles of authority and tutoring in ceremonial matters. "Postreproductive" is clearly a biased term that neglects the many functions of older individuals. Among the !Kung San virtually all of the old folk are storytellers.[9]

Care of the child is a crucial task in foraging societies. An important aspect of the life cycle has to do with the extended circle of family and friends in which a child is reared. The child grows up owning and wanting very little, gaining familiarity with the means and joys of life. Parents are less important in primary care than they are in today's world. Not only are grandparents available but so too are uncles, aunts, cousins, and siblings. And all of these may reinforce each other as they share in the care of the child. As a result the parents themselves may be better friends of the children than in nuclear families where their admonitions, rules, directions, and scoldings might poison a relationship. Lifelong hostility to one's parents is not normal. Perhaps this pathology arises from a deprivation of avuncular and grandparental care.

The old do not sleep through the night as soundly as their grown children—perhaps because their spontaneous wakefulness has been genetically programmed for putting wood on a fire that might go out at 3 A.M., a fire whose flame is not only heat-giving but also a deterrent to predators. If older people wistfully flounder sleepless in our own time, perhaps it is because so many of their adjunct and advisory functions have been lost in the disintegration of the extended family, replaced by the technologies of recollection and decision making, the cult of youth, or their own decrepitude as victims of premature failed health.

A meditative stillness that is good for the human soul, suggests poet Gary Snyder, was invented by motionless hunters.[10] That moment of silent reverence comes also at the final death stroke when one succumbs to the cycle of life.

✦

WHAT CAN WE CONCLUDE from this brief glimpse of the human life cycle? Two things of paramount necessity for proper childhood development come to mind. The first is the opportunity to explore, understand, and become intimately connected to the nonhuman environment that will provide the grounding for symbolic meaning throughout the life cycle. The second is mitigated neoteny: the appropriate social and cultural responses that will guide and support the child, as well as the child in the adult, to his or her final hour. We are given a time plan for our lives—an ontogeny—that is part of our genetic heritage. It commits us to cultural solutions according to a calendar of development. It succeeds only if this social caretaking is in psychological and physical accord with the natural world. If our immaturity is unmitigated we remain stymied throughout our lives, sunk in the symptoms of infantile emotions and demands, of juvenile literalness and materialism, of the violence to which unmitigated adolescent idealism leads as callow ideology engenders dogmatism and impulsive action.

NOTES

1. Lionel Trilling, *Beyond Culture: Essays on Literature and Learning* (New York: Viking, 1965), p. 115.

2. Marina Roseman, "The Social Structure of Sound: The Temiar of Peninsular Malaysia," *Symposium of Comparative Musicology*, Proceedings of the Society for Ethnomusicality, 29th annual meeting, University of California, Los Angeles, 18–21 October, 1984, pp. 414–415. Editor's Note: Louis Sarno and Bernie Krause, in their book *Bayaka: The Extraordinary Music of the Babenzele Pygmies and Sounds of Their Forest Home* (Roslyn, N.Y.: Ellipsis Arts, 1995), describe the Babenzele Pygmies, modern primitives, who are surrounded with music from conception to death in work, in play, and in ceremony. Boys begin drumming on trees, pots, and available objects while they are still toddlers, pluck the bow-harp at seven, and progress to mastery of notched flutes and harp-zithers, which they learn to construct in early adulthood. Girls four years old sing along with the drumming, are improvising at the age of seven, and develop remarkable technical singing skill by the time they are young adults. In a personal communication of 3 October 1997, Bernie Krause explained further: "Sarno (independently) discovered what we found some time ago . . . that where tropical forest-dwelling groups still live closely connected to their environments uninfluenced by Western culture, they tend to use the sounds of their habitat(s) as a kind of natural karaoke orchestra to which they improvise and create their music—something I suspect we used to do very long ago."

3. "The Social Structure of Sound," p. 435.

4. Dolores LaChapelle, *D. H. Lawrence, Future Primitive* (College Station: University of North Texas Press, 1996), pp. 159–160.

5. Geza Teleki, "The Omnivorous Diet and Eclectic Feeding Habits of Chimpanzees in Gombe, Tanzania," in Harding and Teleki, eds., *Omnivorous Primates*, pp. 323–325, notes that in one study forty-eight chimps killed thirty-three colobus monkeys, twenty-three bush pigs, fourteen bushbucks, two redtail monkeys, two blue monkeys, and a baboon in a year. Predation on mammals gained them up to 180 grams per hour, while eating termites gained only 70 grams. Among the chimps the males tended to seek mobile prey, which they surrounded cooperatively and stalked using terrain and plant cover, while the females undertook a "leisurely exploitation of invertebrate resources." Kinship, age, sexual status, and social status, that is, "all the basic behavioral and social forms of human meat-sharing." In hunting they did not use many tools, but then they did not hunt animals as big as themselves.

6. Colin Turnbull, *The Human Cycle* (New York: Simon & Schuster, 1983), p. 50.

7. Harold F. Searles, *The Nonhuman Environment* (New York: International Universities Press, 1960), pp. 100–103.

8. S. Boyd Eaton, Marjorie Shostak, and Melvin Konner, *The Paleolithic Prescription: A Program of Diet & Exercise and a Design for Living* (New York: Harper & Row, 1988), pp. 201–206.

9. Megan Biesele, "Aspects of !Kung Folklore," in Richard B. Lee and Irven

DeVore, eds., *Kalahari Hunters and Gatherers* (Cambridge, Mass.: Harvard University Press, 1976, pp. 302–324).

10. Gary Snyder, quoted in Peter B. Chowka, "The Original Mind of Gary Snyder," *East-West,* June 1977.

IV

How the Mind Once Lived

Our human foraging ancestors who plunged onto the scene in the Pliocene/Pleistocene savannas, emerged from a mammalian ecology that had been under way for around fifty million years. Our particular way of seeing and reading the nonhuman and nonliving world around us evolved out of long-standing strategies of hunting, competing, and evading devised in these first mammals and their evolved prehuman forms. The niche open to these prehumans placed them in a network of open-country mammals whose mental capacities would become fossilized in the shapes and volumes of their craniums as they left their skulls buried in the earth. The human brain, as we progressed toward our human form, and unlike the brain of the prehuman ancestor and its mammalian contemporaries, would double in size (from 500 to 1,000 cubic centimeters from the genus *Australopithecus* to genus *Homo*)—and with this size came the capacity to outsmart other prey and predators on the savanna. Our mind came out of that long-ago scene and we owe its capacity to our ancient ancestors who faced, survived, and adapted to a challenging, rich, wild milieu that remains etched on our craniums like ancient paintings on cave walls.

The "game" we entered was a *trope* with paradigmatic roots in the animals whom we began to hunt and whom we believed to be sentient, intelligent, and spiritual. The game the mammals played among themselves, hunter and hunted on an open playing board, was itself *trophic*, based on energy flow and food chains. This game of predation and survival was a long-standing matter of mutual pruning (over a long period of time, in terms of the minutest statistical advantage, perhaps on the order of 0.0001 percent per century) that favored the most swift, cunning, and discerning over those who were slow to catch on and flee.

All predation is a life-and-death game, but this one was different because open country and eye orientation enabled the thinkers to displace events in both time and space. Imagine, if you will, footprints washing away in a wet jungle, the forest dampening of sound, the hidey-holes that could do for a tactical escape without taxing cognition, and numbers of smaller prey, tiny parcels of protein, in the form of insects, rodents, or reptiles, that would reduce the necessity of a long search or chase. A bent tuft of grass, slowly raising its head, clocks the time since it was trampled; a distant call in known terrain says it is the there, not the here, where attention should be paid. In open terrain, big mobile prey can escape easily and dangerous predators may be forced to strategize from a distance.

A dozen species of large carnivores and an equal number of powerful ungulates played at this game through the veldt and grasslands for sixty million years. As genera they came and went like substitutes on a playing field, losing place as their competitors or their prey upped the intellectual ante. Because of little neural connections, our ancestors were well ahead of most of their kinfolk in the swamps, brush, and forest in terms of discerning the relationships between clues: the color of droppings, the presence of blood, the body language of a pregnant or nursing female, the intentionality of lions, and a thousand other important events that occurred around them.

The game had already started, and we newcomers came to it with our bumptious primate scuffling, our chimpanzee-sized brains, our social preoccupations, and a growing taste for meat. Our ancestors literally walked into this ongoing play in the savannas like naive young things seeking success on the New York stage. But they brought with them the venerable skills of primate scheming, intrigue, and an arboreal and social agility that had characterized simians for millions of years. It did not take us many millions of years to become competitive. There were important advantages we brought along or perfected: bipedality, larger size, the hand, and, of course, that calculating brain already bigger in ratio to body size than most mammals possessed.

At some point learning by imitation was transcended by the unique human capacity to reason abstractly. This ability may have arisen in the course of hunting. Detecting the presence of unseen animals by the calling (or "scolding") of other animals or birds would be such a knack, or reading visual tracks, which is among the most subtle and demanding of human abilities. And there were other clues: the smell of urine, the age and composition of dung, the drying of bitten stems, or the overall pattern of footprints and traces in a day's experience. Escalated into a year's experience, animal migrations could be anticipated by the signs of plant phenology, the phases of the moon and sun, and the changing sounds of the year.

Our ancestors undertook what the chimpanzees never got around to: catching big, dangerous things to eat, things bigger than themselves. The need to cooperate in order to accomplish such a task was probably not apparent at once: it may have grown slowly out of mutual scavenging that required occasional support against competing scavengers or predators returning to claim their kill. This mutual scavenging took place in mixed open country, where the florescence of grasses and their seeds made possible the great herds of ungulates, and they, in turn, entertained a world of carnivores in a slow dance of mental synergism, each taking its cue from the other.[1] Back and forth the predators and prey, including our forebears, tested each other's brains across aeons of successes and failures, always subject to the law that even in evolution you can never do only one thing.

Imagine a dramatic spectacle on a continental scale: a proscenium of grassland and park, in which the players come and go masked as a succession of species, all obedient to the central theme: a banquet at which the participants—eater and eaten—risk the improvements of mind against the certainty of occasional poor decisions, faulty memory, carelessness, errors of judgment, and the decrepitude of age and disease. The overriding rule was simple: the catcher had to be smarter than the caught, but not much and not always. Those who fled had to understand the limits of distance, the intentions of the others, and the ability to control the abyssal terror that itself would engulf them if they submitted to panic. Out of this immense drama among dozens of species of big mammals, herbivore and carnivore, came brains, mind, memory, and strategy—spontaneous and conscious. Those other savanna species who hid in trees or went underground became peripheral, not to the ecology of the whole, but to its mega-chessboard.

Masters of the grasslands and parks, early humans were not. But with a single round of brain cell doubling beyond that of the chimpanzee, they took cognition and communication far enough to find a niche among the faunal cognoscenti like gifted "red-shirt" freshmen making the team. They learned the open-country craft of the hunter and the cunning of the prey, for they were the stalking, tracking predator, the wary, elusive victim, and a passing opportunist in a reciprocity not only between hunter and hunted but within the group and, eventually, aspects of the self.

The less direct consequences of our participation in the game were not just survival skills but the whole panoply of social forms that came to be typical of the primal foraging human groups. Out of their peculiar vulnerability, their proclivity for seeking rock shelters, and their strong primate instincts for communication came the selection of a genome for thinking out events. Group size, growth rates, ontogeny, male/female relations, and the social imperatives of leadership—all were indirectly shaped by the game.

✦

AT THIS POINT in the growth of mind it became fully human and brought into play various complex forms of cognition: the pantomime, the mimicked reference, sharing the idea of an animal by imitating its calls, the way it kicks, stamps, tosses its head, or ritually fights its own kind in stylized performances. A huge repertoire of human communication must surely have grown from this activity, including references made to each other symbolized by types of animals or the pantomimes and gestures in which hands trace forms in the air, keyed to sounds that were not merely mimic. "Sign language" began with conventional signals, just as alphabets began with glyphs. And perhaps out of gesture came the drawn form, just as dance came from the pantomimes. Such gestures were extrapolated even to the sky, so that a constellation or a group of clouds might be "read," or told and come to play their own role in the illustration of narrative. The afterthought was twin to the forethought: from narration of the past, to the articulated plan, to the formulated strategy of the hunt to be.

"Savage thought," says Claude Lévi-Strauss, "is definable both by a consuming symbolic ambition such as humanity has never again seen rivalled, and by scrupulous attention directed entirely towards the concrete."[2] To which W. E. H. Stanner, who has studied Australian Aboriginal history, religion, and ways, adds: "If one wants to see a really brilliant demonstration of deductive thought, one has only to see a blackfellow tracking a wounded kangaroo, and persuade him to say why he interprets given signs in a certain way."[3]

Lévi-Strauss rescued the "savage mind" from the idea that it was childish and stupid. Rather, he identified within it the feature of timelessness, an affinity of all present events with past events. Stanner describes Aboriginal thought as a "metaphysical gift," an idea of the world as an object of contemplation, a lack of omniscient, omnipotent, adjudicating gods, in a world without inverted pride, quarrel with life, moral dualism, rewards of heaven and hell, prophets, saints, grace, or redemption—all this among blackfellows whose "great achievement in social structure," he says was equal in complexity to parliamentary government, a wonderful metaphysics of assent and abidingness, "hopelessly out of place in a world in which the Renaissance has triumphed only to be perverted and in which the products of secular humanism, rationalism and science challenge their own hopes."[4] His contemporaries must have thought Stanner had "gone native" and left his critical intelligence in the outback.

Among primal peoples, observes anthropologist Dorothy Lee, who studied Native American perceptions, there is a "non-linear codification of reality" in which space is not defined by distances on a uniform scale. The lines linking points are not mathematically perceived like the typographic lines in a book; it is a world without tense or causality in language, a world where change is not a measured becoming but a new areness; it is a journey, not a passage through, but a revised at-ness.[5] It is "an event world," signified by sound, says Walter Ong, created from interiors rather than surfaces, returning the hearer always to the organic paradigm, life, the body as the source of sound.[6] It is a world, says Bogert O'Brien about the Inuit, where the transience of objects is their foremost quality and one "does not depend on objects for orientation. One's position in space is fundamentally relational and based upon activity. The clues are not objects of analysis. . . . The relational manner of orienting is a profoundly different way of interpreting space. First . . . the environment is perceived subjectively as dynamic, experiencing processes. . . . The hunter moves as a participant amidst other participants oriented by the action."[7] Humans, as Irenaus Eibl-Eibesfeldt tells us, are hunter/gatherers, irrespective of the cultures that can sometimes obscure and distort what they truly are.[8]

As a special case of this sense of the fluidity of motion in time and place, consider the tradition of running among many Native Americans, a hint of worldwide traditions, mythically and cosmologically integrated, drawn undoubtedly from the esthetics of the chase. Running had "magical ends" and "mystical purposes," including "trance running" or "skimming" in "the hummingbird way." Peter Nabokov describes the "extrasensory perception of the trail" as though it moved under the runner, a special way of "trusting the earth."[9] To spiritualized running one might add nightwalking, which has been explored recently as a way of developing the capacity to see in the dark by training the skills of peripheral vision.[10] Night vision depends on the non-color-sensing (rod cell) parts of the retina surrounding the central (cone-cell) area of keen vision where we focus the images we "look at." By walking at night without looking at the trail (deliberately inhibiting the central area vision) we develop the peripheral field, mediated by the rod cells. Through this exercise we achieve a new level of nocturnal sensibility as well as more acute perceptual abilities. If peripheral information feeds directly into the unconscious, as some believe, we may enhance access to our unconscious by such nocturnal skills as nightwalking. The rational, objective world, which occupies most of us each day, usually overrides the nonrational and unconscious world—which, when neglected, intrudes, disrupts, and overturns our logical mind. In the world of the forager, this was not an issue since the rational and nonrational functions of the brain were balanced and acknowledged. They could see in the dark as well as discern the dark underside of human consciousness.

✦

THE COMPLEX MENTAL PROCESSES involved in foraging are worth looking into in more detail. A sort of venatic phenomenology occurs when primal peoples interpenetrate the nonhuman world in an extraordinary achievement of toolmaking, intellectual sophistication, philosophy, and tradition. As a result, says Lévi Strauss, "in a world where diversity exceeds our mental capacity nothing is impossible in our capacity to become human."

The "savage mind" grasps the world in a totality of present and past with all its multiplicity and complexity. On the other hand, as Lévi-Strauss has revealed, civilized thought attempts to simplify rather than clarify the complexity of the world. It does so by unifying and seeking continuity, variability, and relativity rather than by conceptualizing new schemes, as does "savage" thought, that then become additional objects to be comprehended. Stated simply, the "civilized mind" attempts to simplify and level the world whereas the "savage mind" is not afraid to become enmeshed in its complexity. Birth and death provide the material for a rich and diverse conceptualization [such as initiation ritual] . . . which transcends the distinction between the real and the imaginary." It may seem that primal thought with its spiritual depth is not scientific, but Levi-Strauss regards such thought as "a science of the concrete." He says: "The manner in which primitive peoples conceptualize their world is not merely coherent but the very one demanded where objects are discontinuous and complex." In treating plants and animals as elements of a message, primitive thought discerns "principles of interpretation whose heuristic value" is only recently matched in our society by telecommunications, computers, and electron microscopes and modern information theory. "The entire process of human knowledge," he concludes, "thus assumes the character of a closed system . . . The scientific spirit . . . contributes to legitimize the principle of savage thought and to reestablish it in its rightful place."[11]

A widely shared theme among primal peoples is that of the life of the animal soul. Among prehistoric foragers, as among the !Kung San today, animals were the principal actors in cosmology, a theriomorphic society engaged with humans in a vast play, the theme of which was the reciprocity of killing and renewal, the unity of eater and eaten. The human task was to discover social themes coded in nature and cataloged as taxonomy, told as stories and danced to the rhythms of animal-skin drums.

Memory is central to hunting and gathering, and brain size is directly related to memory. Memory becomes more important the bigger and more dangerous the game, the more helpless and far-traveled the gatherers. Memory is also extremely important to gatherers as they range across home terrain extracting from plants and earth. Moreover, they develop an uncanny vigilance, a softness of presence for which prey species are noteworthy, the ability to become inconspicuous, unnoticeable, such as is seen in a bird collecting food for its nestlings.

Hunting big animals, as noted earlier, requires a kind of timed cooperation that necessitates planning. Richard Borshay Lee's book *The !Kung San* describes the preliminary dialogue and deliberation between hunters before they commence the hunt: the lengthy discussion of the rains, the state of grazing in different localities, the significance of recent sightings. Hundreds of hours per year go into such verbal colloquy and considerations, a discussion that includes women. The dialogue continues to unfold from the preliminary discussions to the hunt itself and after. When tracking, the !Kung San note birdcalls and signs and discuss the spoor. Tracks tell the species, age, sex, speed, and physical condition of the animal and whether it was accompanied by other animals, what it was feeding on, and when it passed. Since tracks change over time, the !Kung San develop "their discriminating powers to the highest degree," estimating how far ahead the animals are. The hunters read the dung and watch for bits of the foliage dropped from the animals' lips while eating. They appraise the size of a herd, whether it has been seeking shade, resting, or halting to feed. The stalking of a wounded animal opens new and repeated discussions and decisions.

During overnight stops the hunters observe specific taboos in a ritually heightened state. Access to the spirits by hunters—ancestral, demonic, plant or animal—is not unusual and can be undertaken in prayer, supplication, dream, trance, visionary disembodiment, and ecstatic flight to the other world. This spiritual state leads to a deeper insight into the meaning of the hunt, the chancy character of the game that may lead to a loss of the hunter's life, and the ethical implications of taking other lives. No hunter on record has bragged that he is master of his fate and captain of his soul. Generally humans have been in the humble position of being few in number, sensitive to the seasons, with an admirable humility, respect, and spiritual connection to the universe. Hunting is, both in an evolutionary sense and individually, says C. H. D. Clarke, "the source of those saving instincts that tell us that we have a responsibility towards the living world."[12]

Storytelling and ritual ceremonies before or after the hunt enhance the spiritual aspect of the hunt. Animal masks in rites give palpable expression to transitional states. On the body of a person the animal mask joins that which is otherwise separate—not only representing human change but conceptualizing shared qualities—so that unity in difference and difference in unity can be conceived as a pervasive truth. And some animals, by their shape or habit, such as foxes and frogs, are also boundary creatures who already signify the threshhold world of human passages. In dance and song, bodies, painted and adorned, move to deep rhythms that bind the world and bring the humans into mimetic participation with other beings and the truth of the multiplicity of all domains.

The most erudite essay on hunting, ancient or modern, is José Ortega y Gasset's *Meditations on Hunting*. His emphasis is on the authenticity of the generic way of being human. He conceives the hunt in terms of a degree of validity of human experience in its direct dealing with the inescapable and formidable necessity of killing. He also refers to the hunter's ability to "be inside" the countryside, by which he means the natural system: "Wind, light, temperature, ground-contour, minerals, vegetation, all play a part; they are not simply there, as they are for the tourist or the botanist, but rather they *function*, they act." Ultimately, this function is balanced by the reciprocity of life and death. Because the mystery of death and that of the animal who comes and goes are the same, "we must seek his company" in the "subtle rite of the hunt." In all other kinds of landscape, he says—field, grove, city, battleground—we see "man travelling within himself" and outside the larger reality.[13] A biologist turned philosopher/historian, Ortega y Gasset regards "primitive" hunter/gatherers as true progenitors of ourselves in the best sense and believes that we realize our true heredity in the hunt.

As noted in earlier chapters, in non-Western, un-industrialized, and largely illiterate (hence nonhistorical) societies, power is plural, societies are egalitarian, and leadership is not monopolized but changing and dispersed. Although there may be said to loom a single creative principle behind it all, in polytheistic worlds there is no omniscience and no single hierarchy, no top-down authority that frames what one is to believe and how one is to live. And still these people have a beautifully fluid yet stable culture that remains intact through climatic and other earthly mishaps. The cement that bonds primal peoples internally and inextricably—the paradigm and exemplar for this social discontinuity among human groups—is the array of natural species about them. Animals and plants are regarded as centers, metaphors, and mentors of the different traits, skills, and roles of people. Insofar as they model diversity and the polythetic cosmos, the animals provide analogs to the multiplicity of stages and forms; they are interlocutors of change that is brought ceremonially into human consciousness.

The foragers' world is rich in signs of a gifting cosmos, a realm of numerous alternatives and generous subsistence, not so much to be controlled by humans as to be understood and affirmed and joined.[14] The original chancy game of prey and predator, of eating or being eaten, takes on a more significant meaning in a gifting world where chance is still an element: the only question is when the gift will pass on. Hunter/gatherers know nature well enough to appreciate how little they know of its complexity. They are engaged in a humble play of adventitious risk, which is hypostasized in gambling, a major leisure-time activity. Gambling is, after all, miniaturizing the game, depicted in the bodies of beasts, lounging or in repose, the ravishing mystery and fun of being a counterplayer, of moving and being moved in the excitement of the chase, the stillness of its sacred aftermath, and the joy of retelling. The great game of chance is elaborated in foragers' myths rich in the strangeness of life with its unexpected boons and encounters, its unanticipated penalties and rewards, not as arbitrary features of supernatural visits but as infinitely complex affiliations.

✦

NOT ALL SEARCHES AND QUESTS are hunts, for the hunt deals with the intense emotional and philosophical problems raised by the act of killing and of facing one's own death. It is not a problem for us simply as predatory carnivores, but as the occasional prey, and as an omnivore whose closest kindred species are also omnivorous, conscious, sentient beings like ourselves. It is right to kill and be killed in this "game" of the hunt so long as we understand the transformations of life and death as a natural consequence of the gifting cosmos where one receives and gives and in the final hour finally passes the gift on. When that clarity is lost the hunt becomes monstrous, along with the rest of nature, and we remove the killing to a butcher's abattoir.

Gathering and hunting are the economic basis of an intricate cosmology in which epiphany and numinous presence are embodied and mediated by wild animals, plants, mountains, and springs. Thinking is toward harmony in a system where people disturb nature so little that its interspecies parities seem to be more influenced by intuition and rites than by physical actions. The hunter's concept of the universe—which Stanner called "the dreaming," also a cosmogony—describes how the universe became a moral system and consists of three elements: marvels, species diversity, and institutions. Marvels refer to that presence of the unexpected that one always encounters sooner or later in nature, particularly when the terrain reflects something about the mind that implies a common structure. His second element, species diversity, coincides with one of the major moral issues of our time—the extinction of species and reduction of biodiversity. Moral issues hinge on real functions, and Stanner has rediscovered what the blackfellow built into his ethical system: that taxonomy is the basic key to human cognition, that thought and speech depend on categories prior to all else, and that morality depends on this as well. If this is so, it makes one wonder how our treatment of each other will change as species are destroyed and diversity is reduced. As for the third element, institutions, they are what humans create most successfully, based on stories of origins, as analogies to the structures that bind the species into marvels of affinity. These are the keys to reality, revealing how things are, what is known, and how to behave.

Tales are a commentary on the underlying principles, a model of morality. Archaeologists Peter J. Ucko and G. W. Dimbleby say: "Some groups of Australian aborigines, despite their extremely limited natural resources and their basically 'Stone Age' technology, have devised one of the most complex of metaphysical systems of belief held by any human group."[15] This cosmogony—how the universe became a moral system—is nothing like an Athenian skeptical philosophy but is a continual, visionary, intuitive, poetic understanding, an ahistoric abiding. There is no quarrel with life. Their metaphysic assents to what men have to be because of the way their life is cast.

The cosmography of tribal peoples is marked by a degree of humility toward the natural world that is lacking in civilized society. Among the principles of the Koyukon worldview, as described by anthropologist Richard Nelson, are these two: "Each animal knows way more than you do" and "The physical environment is spiritual, conscious, and subject to rules of respectful behavior."[16] The worldwide rules of the "sacred hunt" have largely to do with the metaphors that arise as an affirmation of the food-chain structure of nature. Humans are free to create lives and societies according to whatever ideals or fantasies suit them. Huichol yarn paintings of Mexico are visual evocations of stories that integrate the human and nonhuman in dazzling, sophisticated webs, uniquely beautiful works of art and tradition and yet consistent with the cosmologies of Australian Aborigines, African Bushmen, and many tribes of American Indians.

Peoples living at the mind-testing limits of the earth especially express anxiety about the necessary appeasement that is always a part of ceremonies of the hunt. "The greatest danger of life," they say, "lies in the possibility that human food consists completely of 'souls.'"[17] Among the Ainu of Japan, anadromous fishing and shellfish collection are treated culturally like gathering while whalekilling resembles the practices of large-game hunting sustained in myth and ceremony.

Peter Matthiessen writes: "In traditional hunting land and life belong to every member of a community. Greenland's mute sea ice and empty land are not an 'environment' in the Western sense—a human 'habitat' to be exploited. They are the ground of a hard life and the realm of memory and cultural renewal, providing a sense of continuity and tradition which lies at the heart of Inuit well-being. Hunting is the vital nerve of Inuit existence."[18]

✦

EVIDENCE OF A MOTHER GODDESS deity in ancient cultures continues to be of interest in our modern society. Psychoanalysts sometimes argue that sacred maternal imagery corresponds to the visual experience of the newborn. Archaeology offers a variety of feminine objects including the "Venus figures," small, clay, obese, and female, some dating back into the Pleistocene, as evidence of an "original" religion with the female as the central deity.[19] Her bulging body is said to represent pregnancy or prosperity, but the worship of fecundity and superabundance is an agricultural monomania. Advocates of a Great Mother, Earth Mother, or "Lady of the Beasts" argue that she is older than a goddess of agriculture, "bringing culture and manners," but close to "the wild, early nature of humankind," to the "instinct-governed being who lived with the beasts and the free-growing plants." But this notion of the savage's instinct-dominated personality and crude life prior to towns was an eighteenth-century invention, a fantasy of urban disaffection, the civilized idea of prehistory as a nightmare. Belief that primitive people were mentally childlike led psychiatrists, classicists, and others to assert that the goddess "who governs the animal world and dominates instinct and drives, who gathers the beasts beneath her spirit wings," represented a "matriarchal" phase in history. Yet there is no good evidence that our Pleistocene ancestors, although they most likely viewed the earth as mother, worshiped a Great Mother deity in the form of a woman. Such figures emerged with agriculture, and the idealized image of the fecund female was projected onto nature and centered the ego on controlling nature in the form of a governing deity.

The foraging cultures that venerated nature, on the other hand, were radically different from those that replaced them during the past ten thousand years, including those venerating goddesses or gods and thereby denying humanity. Throughout the twentieth century there has been a continuing debate about the meaning of Paleolithic paintings and etchings, primarily of animals in the caves of France and Spain. Speaking of the paintings estimated to be twenty thousand years old discovered in the Grotte Chaivette in France in 1994, Meg Conkey, specialist in Paleolithic art, says: "Cro-Magnon's world was an intensely animated world. . . . They had reindeer in their stomachs, but rhinos on their minds. . . . They were painting the animals that were good to think a bestiary of symbolically important animals."[20] These were not dogs, chickens, or milch cows but wild, free beings who owned the world as much as the hunters themselves, and in whose great beauty *Homo sapiens* had discovered a mirror of the best of human qualities.

Two dozen corpulent, Paleolithic "Venus" statuettes hardly compare to the thousands of animal drawings or the etched figures in stone and abstract signs accompanying them. Everything known about hunter/gatherers of the Pleistocene and the present refutes Levy-Bruhl's first analysis of the Franco-Cantabrian cave art as "fertility magic." The little figurines may signify a collective sensitivity to a quality that could be characterized as feminine, but only as one of many aspects—not as the holy of holies in the figure of a woman. In farming, the womanly representation of feminine productivity became an appropriate model of the generative, nurturing, and renewing processes; but even in Neolithic and Bronze Age cultures of 5,300 to 10,000 years ago she is still part of a mélange of snakes, cattle, and birds. The Great Goddess has much to do with agriculture; but neither she nor agriculture represents the primal, psychologically mature stage in human evolution. She may, however, have become the numinousness of the world through the eyes of a regressive, immature society that had lost the vision of themselves as counterplayers in a vast cosmos of other species or "peoples," a society that had become instead the caretakers of seeds and livestock much as they themselves had been nurtured by their mothers.

✦

THE COMPLEX WEB OF SKILLS and knowledge of the hunters and gatherers is translated into a fluid design of social interaction as well as an all-encompassing cosmology that speaks to a rich spiritual life lived in the shadow of death. Hunters are engaged in a game of chance amid heterogeneous, exemplary powers rather than collective strategies of accumulation and control. They never play it as "maximizing your take." Their metaphysics conceives a living, sentient, and diverse comity whose main features are given in narrations that are outside History. Their mood is assent and affirmation of their circumstances. Their lives are committed to the understanding of a vast semiosis, presented to them on every side, in which they are not only readers but members. The hunt becomes a kind of search gestalt. The lifelong test and theme is "learning to give away" what was a gift received in the first place—life itself—a theme demonstrated daily in the sharing of meat.

NOTES

1. Harry J. Jerison, *The Evolution of the Brain and Intelligence* (New York: Academic Press, 1973).

2. Claude Lévi-Strauss, *The Savage Mind* (Chicago: University of Chicago Press, 1966), p. 220.

3. W. E. H. Stanner, *White Man Got No Dreaming* (Canberra: Australian National University Press, 1979).

4. Ibid.

5. Dorothy Lee, "Lineal and Non-Lineal Codifications of Reality," *Psychosomatic Medicine* 12(2) (1950): 89–97.

6. Walter J. Ong, "World as View and World as Event," *American Anthropologist* 7(4) (1969): 634–647.

7. Bogert O'Brien, "Inuit Ways and the Transformation of Canadian Theology," unpublished manuscript, 1979.

8. Irenaus Eibl-Eibesfeldt, *Love and Hate* (New York: Aldine de Gruyter, 1996).

9. Peter Nabokov, *Indian Running* (Santa Barbara: Capra, 1981), p. 144.

10. Nelson Zink and Stephen Parks, "Nightwalking: Exploring the Dark with Peripheral Vision, *Whole Earth Review*, Fall 1991, pp. 4–9.

11. Lévi-Strauss, *The Savage Mind*, pp. 263–268.

12. C. H. D. Clarke, "Venator the Hunter," unpublished manuscript, n.d.

13. José Ortega y Gasset, *Meditations on Hunting* (New York: Scribner's, 1972), pp. 141–142.

14. Nurit Bird-David, "The Giving Environment," *Current Anthropology* 31 (2) (1990): 189–196.

15. Peter J. Ucko and G. W. Dimbleby, "Introduction," in *The Domestication and Exploitation of Plants and Animals* (Chicago: Aldine, 1969), p. xviii.

16. Richard Nelson, *Make Prayers to the Raven: A Koyukon View of the Northern Forest* (Chicago: University of Chicago Press, 1983), p. 225–231.

17. See Wilhelm Dupre, *Religion in Primitive Cultures: A Study in Ethnophilosophy* (The Hague: Mouton, 1975), p. 206, referring to Kaj Birket-Smith, *Die Eskimos* (Zurich, 1948).

18. Peter Matthiessen, "Survival of the Hunter," *New Yorker*, 24 April 1995, pp. 67–77.

19. Marija Gimbutus, *Goddesses and Gods of Old Europe* (Berkeley: University of California Press, 1982).

20. Meg Conkey quoted by Virginia Morell, "Stone Age Menagerie," *Audubon*, May–June 1995, p. 54.

V

Savages Again

After twenty centuries of ideological controversy about "savages" and the lost primal world it may be impossible for us to look at the applicability of Pleistocene lifestyle options for the modern world without trailing biases, illusions, and romanticism about tribal people. The historical image of the savage suffers from two extreme views: the paragon of the noble savage at one limit, the loathsome brute at the other. The first is supported by a mythology of Golden Age legends. The second draws its energy from the history of cultural chauvinism—the idea of "savage" degradation and its "animalistic" expressions.

The idea that our primal ancestors had only a fuzzy and passive self-consciousness has haunted anthropology and history for most of the twentieth century. "Primitive man," said Jane Ellen Harrison in 1912, "submerged in his own reactions and activities, does not clearly distinguish himself as subject from the objects to which he reacts, and therefore has but slight consciousness of his own separate soul."[1] Manuel Navarro, as late as 1924, said of the South American Campa: voracious brutes, "degraded and ignorant beings, they lead a life exotic, purely animal, savage, in which are eclipsed the faint glimmerings of their reason, in which are drowned the weak pangs of their conscience, and all the instincts and lusts of animal existence alone float and are reflected."[2] Closer to home is the testimony of Will Durant, the historian: "Through 97 percent of history, man lived by hunting and nomadic pasturage. During those 975,000 years his basic character was formed—to greedy acquisitiveness, violent pugnacity and lawless sexuality."[3]

"Humanist anthropologists" like Edward Tylor and Jacob Malinowski dismissed native religious rites as logical error but allowed that ritual might work psychologically. Although there have been bold voices among them, such as Marshal Sahlins' *Stone Age Economics*, few anthropologists advocate a new primitivism. Their restraint is the result of a hard-won professional objectivity, the effort to overcome centuries of ethnocentrism, along with the pressures of cultural relativism in the social sciences, pioneered by Franz Boas and Alfred Kroeber.[4] As to the veracity of primal religions, an embarrassed silence has marked anthropology ever since.

Robert Edgerton presents a menu of rotten behavior from one or another small society: rape, homicide, genital mutilation, wife beating, torture, child abuse, infanticide, and other antisocial traditions and beliefs. He claims that harmonious small societies never did exist and that the notion of primitive societies with more humane and kinder practices, a better ecology, less pathology, and a spiritual life in keeping with life on Planet Earth is a romantic fantasy. His colleagues, Edgerton says, failed to record the dark side of small societies. Their misleading reports are part of anthropology's "relativist" attitude—no judgment of cultures other than their own.[5] Along with tribal-level farmers, he cites sedentary potlatch fishermen and domestic-ungulate-focused pastoralists whose lives are shaped by competition, ownership, and power alliances.[6] He does not acknowledge the psychopathic consequences of human density, the scale of suffering, or its wider environmental effects in mass societies. He speaks of "opposing interests" and "competing interests" among hunter/gatherers as though these were intrinsically destructive. But inequality is not necessarily bad: variations in size, strength, sex, and temperament are the basis of all animal societies.

Edgerton's examples are limited to habitat fringe groups: the Inuit, the Papuans, and the Siriono of Bolivia, for example. The Inuit and Papuans live at environmental extremes in perpetual snow or tropical forests—far from tundra, steppe, savanna, and forest edge in which *our ancestors* forged the kind of species we are. The Papuans and Inuit may be nearly as far from their evolutionary home as we are in cities. "Maladaptive" behavior is no surprise. All such small island, wet forest, and high arctic societies are deep into fending off evil and dangerous spirits.

The idea of the inherent "nobility" of the individual savage was laughed out of school a century ago, and properly so. Foragers are not always pacific (though they do not keep standing armies or make organized war), nor are they innocent of ordinary human vices and violence. In one or another group there is small-scale cruelty, infanticide, and inability or unwillingness to end intratribal scuffling or intertribal vengeance.

Edgerton does allow that our Pleistocene genetic heritage is maladaptive in post-Pleistocene environments. Our craving for salt, sugar, and fat, for instance, is healthy in wild environments where salt is not normally superabundant, sugar stimulates the appetite for fresh fruit, and saturated fats are limited and wonderfully balanced in wild animals. Our built environments, moreover, may also be maladaptive as well as unhealthful. Are environments of stone and concrete psychically toxic, or is it having more than a few dozen people around at once that creates our unexplainable syndromes?

Edgerton knows that relativism based on multiculturalism only goes so far. Human psychobiology, he admits, is the same everywhere (but not the cultural response to it). There are universal human needs and characteristics, many of which are positive. We inherit, he says, such things as a predisposition for ways of life that are nomadic, we divide tasks by gender, and favor social arrangements that are typically sharing and mutually supportive. This perspective stands in sharp contrast to the Freudian psychocentric notion that "primitives" are like children or Kroeber's view of them as psychopathic.

Contrary to the skepticism about primitive cultures, perspectives from various quarters—the study of higher primates, hominid paleontology, Paleolithic archaeology, ethology, ecology, field studies of living hunter/gatherers, direct testimony from living hunter/gatherers — provide powerful examples for thoughtful speculation about our ancient ancestors. Revelations come from the meticulous ransacking of old campfire sites, the artifacts of ceremony and the ensemble of art as the tangible evidence of mind, analogies to the eating strategies of other species, and the social and ecological homologies with living foraging peoples.

✦

HOSTILITY TO THE IDEA that we have anything to learn from savages has a long tradition. For two centuries the ideology of progress set its values opposed to fictional images of deprived and depraved savages. It is the whole of personal existence, from birth through death, among those whom history calls "preagricultural" peoples that is the strength of a Pleistocene way of life.

Much is to be learned from today's hunter/gatherers despite the fact that contemporary hunters are not our ancestors.[7] Although we cannot declare that past cultures are repeated unequivocally in the present, we can assume that there are similarities between peoples whose lifestyle is comparable whether they be archaic or modern foragers. After the proceedings of a Wenner-Gren symposium in Chicago were published as *Man the Hunter* in 1968, it was clear that categorizing primitive humans as either brutish cavemen or noble savages was an erroneous interpretation of the complexity of original culture. Field-workers who had studied living tribal peoples in many parts of the world came together and found common threads that linked diverse hunter/gatherer cultures to one another and to Paleolithic archaeology.[8]

All people instinctively differentiate among themselves socially. But cultures differ in their criteria for doing so. Among hunter/gatherers the criteria tend to be age, gender, and ability; in complex societies the social distinctions are more often wealth, power, and kingship. In the bosom of family and society, the life cycle is punctuated by formal social recognition with its metaphors in the terrain and the sentient plant and animal life surrounding the human community.

Typical characteristics of family-level economy, say Allen Johnson and Timothy Earle, are low population density, personal tools, familiar seasonal rhythms of aggregation and dispersion, nonterritoriality, no war of professional soldiers or standing armies, familiar ceremonialism, an ad hoc leadership, and much personal choice, including risk taking.[9] In the last two centuries individuals in primal societies have been given short shrift and depicted as sacrifices to mass impulse, un-self-reflective organisms who are utterly dependent on the tribe. This supposed lack of self-consciousness has given us a picture in the hands of classical authors and modern psychiatrists of some defective state in human evolution that awaited civilization. According to anthropologist Elman Service, however, "an individual adult participates much more fully in every aspect of the culture than do the people of more complicated societies. . . . Human beings in primitive society are personalized and individuated."[10] Among Ituri pygmies, "outrageously boastful men and extremely shrewd women," who show "humor, gaiety, reflectiveness," all "contradict the conventional image of preliterate peoples as divested of ego and personality."[11] comments sociologist Murray Bookchin.

One leads by assent of others—by listening, arguing, suggesting, and reflecting a consensus, a spontaneous accord that always has limits. "The open competition of leadership in an Indian community," says geneticist James V. Neel, "probably results in leadership being based far less on accidents of birth and far more on innate characteristics than in our culture."[12] Anthropology itself has contributed to a picture of wild men in a bloody melee. "Because we read so much about animism and magic, totemism and demons, we come to identify primitive people with these things unintentionally and to imagine them as always plagued by demons, or running into taboos, and passing their lives in a chronic state of terror," observes Geza Roheim.[13]

Interpersonal and gender relations are worked out in the context of daily life of hunting/gathering and sharing, where leadership does not take on special significance. Colin Turnbull says: "In terms of a conscious dedication to human relationships that are both affective and effective, the primitive is ahead of us all the way."[14]

Group size in foraging groups is ideal for human relationships—including vernacular roles for men and women without sexual exploitation. The idea of a vernacular gender was widely misunderstood in the antagonistic atmosphere of the 1980s, in the anger that repudiated four thousand years of male sovereignty.[15] Men and women are unlike because of their evolution, a matter not to be deplored but to be celebrated and fulfilled, with the caution that power over the other is not part of the difference. Roles and duties are divided, but not to make inequality. A vernacular society, divided in many of its social and familial responsibilities and privileges, would be inappropriately dominated by either gender. Men and women have different roles in the group, similar and yet different bodies and psyches, shared but also different satisfactions, desires, fears, and sorrows. In small-group societies such a complementarity is both beautiful and efficient. Yet diversity of sexual orientation or social role is respected as well. In such societies those who are ambiguous socially or sexually, who do not marry and have children, are not penalized but occupy legitimate roles. Marjorie Shostak in her study of the !Kung San asserts that although "men's status is sometimes higher than women's, still it must be said that "women have considerable voice in group affairs and considerable control over their own lives (that is, in terminating an unsatisfactory marriage). In these respects they may be more egalitarian than most other societies, including our own."[16]

Health is good among the !Kung San in terms of diet as well as social relationships.[17] They eat 80 of the 262 species of animals they know, but with no effect on the animal populations. James V. Neel observes "The high level of maternally derived antibodies, early exposure to pathogens, the prolonged period of lactation, and the generally excellent nutritional status of the child make it possible for a *relatively* smooth transition from passive to active immunity to many of the agents of disease to which he is exposed."[18]

How much do foragers work? "No group on earth has more leisure time than hunters and gatherers, who spend it on games, conversation and relaxing," according to Frank Hole and Kent V. Flannery.[19] Among the Nunamiut Eskimos, "Umialit (rich men) were considered very intelligent men who carefully observed the habits of all the animals and the conditions affecting them: climatic, topographic, other animals, and the presence of men."[20] The "fashionable, male-oriented hunting models," according to Glynn Isaac and Diana Crader, have little to do with true foraging cultures. The taking of wild meat and its use, particularly, involve a division of labor, food sharing, information on the past and future, long-term mating bonds, joint care of the offspring, "networks of reciprocity, spiraling developments in communications, and intragroup cooperation."[21]

The beauty of Pleistocene "work" is that as such it hardly exists in the sense of the modern concept of labor and hourly drudgery. The work week is about seventeen hours, and although carrying meat or wood to camp may have its tedious moments, most of hunting and gathering activities, as well as dancing and games, exercise those muscle and coordination complexes that we now see as beneficial exercise. Running is particularly evident.

Among foragers the esteem gained in sharing and giving outweighs the advantages of hoarding. According to Richard Borshay Lee, "there is no evidence for exploitation on the basis of sex or age" and sharing (other than sex) is the most singular ideal and obligation.[22] The only private property is personally constructed things. The worst accusations are stinginess and browbeating. Among hunters and gatherers custom firmly modulates human frailty and irascibility. Fights are more likely to be over sex, adultery, and betrothal than land and resources. Land "ownership" is a collective understanding. Outsiders are not excluded. Mobility allows for easy dispersal and joining, splitting and coalescing, for social or ecological reasons. Organized war does not exist. Ecological affinities are stable and nonpolluting. There is, says Lee, "a continuous struggle against one's own selfish, arrogant, and antisocial impulses. . . . A sharing way of life is not only possible but has actually existed in many parts of the world and over long periods of time."

People living in primitive societies do not seek to move to higher human density situations, but instead move to lower-density areas where resources are more abundant. As Johnson and Earle have shown with the Michiguenga in Peru, "as long as wild foods remained abundant, the village succeeded in handling the numerous tensions of group life." When resources became scarce, however, tensions arose "and disputes erupted into open conflicts" that were resolved by moving the whole village to a new site.[23]

When on rare occasions "there is disagreement over hunting plans, it is usually resolved by making one's views known to all and reaching an acceptable consensus of opinion through public discussions participated in by both males and females," says Susan Kent of the Bushmen.[24] Lee explains that differences in skills do not necessarily create friction: "The correlation between hunting success and social status is minimal and tends not to be emphasized." There is much variation among the men in hunting skill—in fact, one-third of them kill two-thirds of the game. There is no traditional way for a man to be a nonhunter, but an individual may not hunt for many months. If there are slim days the !Kung San do not go hungry for long, compared to the northern Ghana farmers. Says Lee: "There is still no evidence for a weight loss . . . even remotely approaching the magnitude of loss observed among agriculturists."[25]

Hunting is associated with an equitable division of labor between men and women: food sharing within the extended family and even wider information sharing about daily experience and the tribal past.[26] Hunting has never excluded women. Their lives are as absorbed in the encounter with animals, alive and dead, as are the men's. The hunt is a continuum that includes the entire community, from its first plan to its storied retelling, from the social analogies to the behavior of carnivores to metaphors on food chains to prayers of apology and thanksgiving. Traditionally the large, dangerous mammals are usually hunted by men, but it has never been claimed that women only pluck and men only kill.

The centrality of meat, the sentience and spiritual source from whom it comes, the diverse activities in its preparation and distribution, the animal's numinous presence after its death—all entail a wide range of roles, many of which are genderized. Insofar as the animal eaten is available because it has learned "to give away," there is no more virtue in the actual chase or killing than the transformation of its skin into a garment, the burying of its bones, the drumming that sustains the whole group as dancer of the mythical hunt, or the dandling of infants as the story of the hunt is told. Women sing the spirit of the slain animal a welcome to the hearth where she is the hostess.

Among the Sharanahua of South America, when the women are meat-hungry they send the men off to hunt and sing the hunters to their task. They are commonly said to transform boys into hunters. Anthropologist Janet Siskind says: "The social pressure of the special hunt, the line of women painted and waiting, makes young men try hard to succeed."[27] Gathering, like hunting, is a lighthearted affair done by both men and women. The stable sexual politics of the Sharanahua, "based on mutual social and economic dependence, allows for the open expression of hostility," a combination of solidarity and antagonism that "prevents the households from becoming tightly closed units."[28]

Since a critical dimension of the hunt is the confrontation with death and the incorporation of substance in new life, women are traditionally regarded as keepers of the mystery of death-as-the-genetrix-of-life in all expressions of sharing and giving away. The hunt is clearly connected with feminine secrets and powers—with the Greek goddess Artemis, for example, and her other avatars such as the archaic "Lady of the Beasts." Paleolithic female figurines occur in sanctuaries where the walls are painted with hunted game.

More value is placed on men than women only as the hunt is perverted by sexism and training for war. Among the !Kung San the women collect small animals but do not hunt. Their nonhunting is not an issue, however, nor an area of abrasion between the sexes. Perhaps sexism comes into being with the doting on fertility and fecundity in agriculture and the androgynous "reply" of nomadic, male-dominated societies of pastoralism.

The metaphysics of meat embraces a range of activity. Excessive genderizing about meat-eating is a popular issue. Human carnivory takes nothing from foraging women who, like people everywhere, prefer meat. In the anthropology of the 1980s and 1990s there has been a continuing discussion of primitive diets—in part because of the general public repudiation of "red meat," which has been shown to contribute to high levels of cholesterol. Not all "red meat" is high in cholesterol, however. In contrast to domestic animals, the fats in wild animals, including seal and walrus, are unsaturated.[29]

According to Lee, the better hunters do not "dominate politics" or play the role of "big man." Modesty and understatement are admired and there is communal pressure to be a good, generous hunter. There is no clear focus of authority or enforcement. The participation of women in group politics is "greater than that of women in most tribal, peasant and industrial societies." Meat is distributed according to kinship and friendships and the circumstances of the kill; even the maker of the arrow is important.[30] Bookchin comments: "the most important attributes of citizenship derive more directly from the tribal world than the village world, from rude shelters rather than houses . . . [towns subvert] the conditions for an active, participatory body politic."[31]

Something enormously powerful binds living hunter/gatherers to all those of the past and to modern sportsmen, who are no exception to the best traditions of the ancient hunt. That something is the way the hunt satisfies the demands of the genome. Hunting is a kind of cross-cultural theme. Metaphorically understood, the hunt refers to the larger quest for the way: the pursuit of meaning and contact with a sentient part of the environment and the intuition that nature is a language. In a sense hunting is a special case of gathering.

Antihunters, including many academics and historians, are too quick to accuse hunters of brutality and to cite naturalists like Aldo Leopold, George Bird Grinnell, William Temple Hornaday, or Henry Thoreau as nonhunters or at least reformed hunters. Thomas Altherr and John Reiger have rightly taken the intellectuals to task for their poor scholarship in these matters.[32]

Hunting, moreover, is sometimes confused with war because many suppose that weapons and aggression are the crux of the hunt. The repugnance often expressed toward hunting is emotionally inextricable from the horror of war. The innocence of the hunted animals, the use of weapons of war against those without weapons, the seeming vainglory in the "trophy" hunt, the apparent infliction of unnecessary pain, the associated atavisms such as violence, aggression, and ferocity, the human adoption of the model of the dire carnivore, the association with commercialization and intoxication—all these and more weigh against the hunt as a falling away from things civilized. The difficulty is that, although there is truth in these criticisms today, the analogy of the hunt to warfare and crime is deeply wrong. Hunting and gathering interplay mind and attention with logic and compassion for the land, forests, water, plants, and animals. Modern sportsmanship and its ethics of the hunter's voluntary restraint constitute a last barrier against the corruption of hunting.

The sense of sanctity and perfection with which primal people glow is a reflection of something essential in human nature. It has to do with an insight about the world—vouchsafed to us all but realized in hunting/gathering cultures—obscured by the inroads of all other ways of life. There is an ineluctable way of being human on this planet, in a world of others, where the flow of life is also the flow of death. Susan Kent says that in "groups without domestic animals, both human and non-human animals are viewed as having an intellect—that is, sentience, sociability and intelligence—and a common mythical ancestry with humans," a kinship that is not shared with plants although the latter may have ritual significance.[33]

"Hunters" is an appropriate term for a society in which meat, the best of foods, signifies the gift of life incarnate—the obtaining and preparation of which ritualizes the encounter of life and death. The human kinship with animals is faced in its ambiguity, and the quest of all elusive things is experienced as the hunt's most emphatic metaphor.

Mistakes about this mode of life hound even anthropologists, some of whom confuse predation with hunting and see hunters only as food extractors.[34] The contrary is true. Foraging peoples typically spend thousands of hours every year pondering and studying the animals around them and discussing the events of the day. The animals are numinous and oracular signifiers. In their most subtle moves, they are watched and studied with dedicated determination.

JOSEPH CAMPBELL HAS ARGUED, rightly, that death was a metaphysical problem for the original hunter. He concluded, wrongly, that it was solved by planters who made sacrifices to forces governing the annual sprouting of grain. It was control, not acquiescence to this great round, that the agriculturists sought. In the dawn of the modern world, the Neolithic, says Wilhelm Dupre, "the individual no longer stands as a whole vis-à-vis the life-community in the sense that the latter finds its realization through a total integration of the individual—as is the case by and large under the conditions of a gathering and hunting economy."[35]

Unlike agrarians and pastoralists, foragers do not perceive nature as simply a larder in which the animals are mere objects in a game of power and wealth. It would be wrong to see this play as a ravagement. Subtlety, restraint, cogitation, and cooperation are its guiding principles. Ferocity has its place, not as a melody, but as a chord. The beleaguered modern tycoon who says of his work, "It's a jungle out there," is in error about the real jungle. His metaphor is a self-serving misrepresentation of the wilderness that made him possible.

From the time of Vasco da Gama, Westerners have been fascinated by indigenous punishment for crimes and by cannibalism (although cannibalism is primarily a trait of horticulturists). Being subject to ordinary human shortcomings, hunter/gatherers may not always live in perfect harmony with nature or each other. Nor are they always happy, content, well fed, free of disease, or profoundly philosophical. Like people everywhere they are, in some sense, incompetent.[36]

Melvin Konner, Harvard-bred anthropologist, who spent years studying the !Kung San of the Kalahari Desert of Africa, wrote of the superiority of their lives to their counterparts in Cleveland or Manchester—and then pulled the covers over his head by saying, "But here is the bad news. You can't go back."[37] One can only be grateful for Loren Eiseley and Laurens Van der Post, writing on the same Kalahari Bushmen, and for their anticipation of what Roger Keesing calls a "new ethnography" that seeks "universal cultural design" based on psychological approaches.[38] "If a cognitive anthropology is to be productive," he says, "we will need to seek underlying processes and rules." He concludes that "the assumption of radical diversity in cultures can no longer be sustained by linguistics."[39] Which is to say that linguistic differences are merely one of the freedoms made possible by the genome.

We are free to create culture—and have done so in hundreds of ways—but there is a catch. The biological function of culture is probably the versatility that it offers to a traveling species, whose environment differs widely and whose experiences are diversely assimilated and built upon, and who need to keep their sense of identity. For thousands of years culture helped set small groups of people apart from each other by embedding their customs and skills and by semi-isolating linguistic and genetic groups. The catch is that, given a natural world and a human nature, not all cultures work equally well. The most rewarding theme was that it was the small-group foraging people who developed the general human niche during the evolution of the genome—the genome which in turn would expect just that sort of small group.

Notes

1. Jane Ellen Harrison, *Themis* (New York: University Books, 1962), p. 475.

2. Manuel Navarro, *La Tribu Campa* (Lima, 1924), quoted in Gerald Weiss, "Campa Cosmology," *American Museum of Natural History Anthropological Papers* 52, pt. 5 (New York: American Museum of Natural History, 1975).

3. Dr. Will Durant, "A Last Testament to Youth," *Columbia Dispatch Magazine*, Feb. 8, 1970.

4. Derek Freeman, letter, *Current Anthropology*, October 1973, p. 379.

5. Robert B. Edgerton, *Sick Societies: Challenging the Myth of Primitive Harmony* (New York: Free Press, 1992), pp. 1–104.

6. Ibid., p. 88.

7. Sherwood L. Washburn, ed., *The Social Life of Early Man* (New York: Wenner Gren, 1961); G. P. Murdock, *Ethnographic Atlas for New World Societies* (Pittsburgh: University of Pittsburgh Press, 1967). The shift toward species-specific thinking benefited from "the new systematics," an evolutionary perspective from genetics and natural selection articulated by George G. Simpson, Ernst Myer, Theodosius Dobzhansky, Julian Huxley, and others. Next came *The Social Life of Early Man*, the new discovery of continuity among primitive societies, later given cross-cultural generalizations in George Murdock's ethnographic atlas.

8. Richard B. Lee and Irven DeVore, eds., *Man the Hunter* (Chicago: Aldine, 1968).

9. Allen W. Johnson and Timothy Earle, *The Evolution of Human Societies from Foraging Group to Agrarian State* (Palo Alto: Stanford University Press, 1987), 28–61.

10. Elman R. Service, *The Hunters*, 2nd ed. (Englewood Cliffs: Prentice-Hall, 1979), p. 83.

11. Murray Bookchin, *The Rise of Urbanization and the Decline of Citizenship* (San Francisco: Sierra Club Books, 1987), p. 27.

12. James V. Neel, "Lessons from a 'Primitive' People," *Science* 170 (3960) (20 November 1970): 818.

13. Geza Roheim, *Children of the Desert* (New York: Basic Books, 1974).

14. Colin Turnbull, *The Human Cycle* (New York: Simon & Schuster, 1983), p. 21.

15. Ivan Illich, *Gender* (New York: Pantheon, 1982).

16. Marjorie Shostak, "A !Kung Woman's Memories of Childhood," in Richard B. Lee and Irven DeVore, eds., *Kalahari Hunters and Gatherers* (Cambridge, Mass: Harvard University Press, 1976), p. 277

17. Gina Bari Kolata, "!Kung Hunter-Gatherers: Feminism, Diet, and Birth Control," *Science* 195 (4276) (28 January 1977): 382–338.

18. Neel, "Lessons from a 'Primitive' People," p. 819.

19. Frank Hole and Kent V. Flannery, "The Prehistory of Southwestern Iran: A Preliminary Report," *Proceedings of the Prehistoric Society* 33 (1963): 201.

20. Michael Jochim, *Hunter-Gatherer Subsistence and Settlement: A Predictive Model* (New York: Academic Press, 1976), p. 22, quoting N. J. Gubser, *The Nunamiut Eskimos: Hunters of Caribou* (New Haven: Yale University Press, 1965).

21. Glynn L. I. Isaac and Diana C. Crader, "To What Extent Were Early Hominids Carnivorous? An Archaeological Perspective," in Harding and Teleki, eds., *Omnivorous Primates*, pp. 37–103.

22. Richard Borshay Lee, *The !Kung San* (New York: Cambridge University Press, 1979), pp. 459–460.

23. Johnson and Earle, *From Foraging Group to Agrarian State*, pp. 82.

24. Susan Kent, "Cross-Cultural Perceptions of Farmers and Hunters and the Value of Meet," in Susan Kent, ed., *Farmers as Hunters* (London: Cambridge University Press, 1989), p. 4.

25. Lee, *!Kung San*, p. 296.

26. Ibid, p. 437.

27. Janet Siskind, *To Hunt in the Morning* (New York: Oxford University Press, 1973), p. 101.

28. Ibid., p. 109.

29. Alan E. Mann, "Diet and Evolution," in Harding and Teleki, eds., *Omnivorous Primates*, p. 23.

30. Lee, *!Kung San*, pp. 343–346.

31. Bookchin, *The Rise of Urbanization* , p. 28.

32. Thomas L. Altherr and John F. Reiger, "Academic Historians and Hunting: A Call for More and Better Scholarship," *Environmental History Review* 19(3) (Fall 1995). Besides Ortega y Gasset's classic statement on the ethics and cosmology of the hunt there are excellent treatises such as James A. Swan's beautiful book, *In Defense of Hunting* (San Francisco: Harper, 1995); Charles Clifton's "Hunter's Eucharist," *Gnosis* Fall 1993; and Matt Cartmill's *A View to Death in the Morning* (Cambridge, Mass: Harvard University Press, 1993).

33. Kent, "Cross-Cultural Perceptions," p. 11.

34. Tim Ingold, *The Appropriation of Nature: Essays on Human Ecology and Social Relations* (Iowa City: University of Iowa Press, 1987).

35. Wilhelm Dupre, *Religion in Primitive Cultures* (The Hague: Mouton, 1975), pp. 327–328.

36. In the movie *Little Big Man* the Indian actor Dan George did an unforgettable satire on the wise old chief who, delivering his rhetoric of joining the Great Spirit, lies down on the mountain to die and gets only rain in the face for his trouble.

37. Melvin Konner, *The Tangled Wing: Biological Constraints on the Human Spirit* (New York: Holt, 1982).

38. Loren Eiseley, "Man of the Future," in *The Immense Journey* (New York: Random House, 1959). Laurens van der Post, *The Heart of the Hunter* (New York: Harcourt Brace Jovanovich, 1980).

39. Roger M. Keesing, "Paradigms Lost: The New Ethnography and New Linguistics," *Southwest Journal of Anthropology* 28 (1972): 299–332.

VI

Romancing the Potato

THE TRANSFORMATION from hunter/gatherer to agrarian economies took place over the past twelve thousand years. This length of time is insignificant in terms of geological history—or, for that matter, in terms of human history that began with the appearance of *Homo sapiens* some four hundred thousand years ago, our genus, *Homo*, at two million years, and our family, Hominidae, six million years ago. Accompanying changes in the face of the land and lifestyle of the people was a concomitant alteration in perceptions of the agrarian participants. The game of comity of life and death, which the hunter/gatherers entered in the great savannas, accepting the nature of nature, was altered by agrarian thought: from a core process of chance to one of manipulation, from reading one's state of grace in terms of the success of the hunt to bartering for it, from finding to making, from sacrament received to negotiations with humanlike deities. The transformation took place slowly and for various reasons, but the result was to concentrate populations in certain areas and make them dependent on the products of domestication.

Between about twelve thousand and eight thousand years ago this transformation in human culture took place in the eastern Mediterranean and Near East. We begin with small, semimobile groups living in what we would now call "wilderness," upon which their impact was small. Then, here and there, little patches of wheat grasses, intensified monitoring of some wild goats or sheep, and the hangdog shadows of scavenging wolves whose offspring were sometimes captured and tamed, all made little pockets of the first agriculture. The topography of ancient Mesopotamia, composed of arid lowlands, mountains, and aggrading streams whose gravel bars were the homes of annual plants in different altitudinal zones, had already resulted in different human economies. The details of the first agriculture are still being debated, but the outlines seem clear. Seminomadic hunter/gatherers in this part of the world had long since seen the last of the elephant, hippo, and rhino. Before twelve thousand years ago the elk, reindeer, horse, and great auroch were disappearing because of climatic changes. A trend in foraging was toward crabs, clams, turtles, fish, snails, waterfowl, and the cereal plants.

The first domestic plants and animals were wheat, barley, goats, sheep, and dogs. Humans have been around thirty-three times as long as the dog. Domesticated cattle are recorded at nine thousand years ago, and horses at six thousand. Almost any typical wild species for which there are fossils are hundreds of thousands of years old. From an evolutionary and geological perspective, the animals and plants that share our homes and our fields came into our lives only yesterday and exist because of the protective care we have given them.

Stones, the first tools of agriculture, originally used for grinding gathered seeds or ochre for body painting, became important implements for grinding harvested grains, and flint sickles were used for harvesting. Wild species diversity diminished. The seed heads of the grasses were selectively modified for storage and planting. Sheep, gazelle, and onager were driven and penned. Planting, storing, and keeping caprine animals and bovines spread from upper grassy slopes to intermontane plains and marshy areas. Irrigation made its appearance in the lowlands. Life was no better for humans than it had been, but the economy demanded more people to reshape production.

Domestication changed means of production, altered social relationships, and increased environmental destruction. From ecosystems at dynamic equilibrium ten thousand years ago the farmers created subsystems with pests and weeds by the time of the first walled towns five thousand years ago.[1] At least six millennia of mixed tending and foraging followed the earliest domestications, preceding the wheel, writing, sewers, and armies. In varying degrees primal foraging blended with early farming. Before cities, the world remained rich, fresh, and partly wild around the little gardens and goat pens. Extended family and small-scale life incorporating the rhythms of the world made this "hamlet society" humane and ecological. Village horticulture, relatively free of commerce and outside control, may have been an ideal life.

Keeping the hoofed animals out of the seed patches and guarding stored food reduced human mobility. The trampling of human feet and hooves around home sites, the progressive use of local wood for fuel and construction, and the accumulation of implements too bulky to carry were among the first material signs of hamlet life and domestication. Fleas, tapeworms, and other parasites were acquired from, and shared with, kept animals. Modification of the surrounding plants into "pioneer" or weed communities simplified and destabilized the environment. As the techniques for storing and corralling became part of the cultural skills, cattle and vegetables were added. Fences made their appearance, and domestic plants and animals created a new company of altered forms.

Wild things retreated into the distance, and the mix of garden, pasture, dwellings, weeds, kept animals, lice, cockroaches, bedbugs, house mice, rats, and other inhabitants of simplified communities filled the phenomenal and economic world. With irrigation, cultivation, and the rest of the routine round of obligatory labor, the human environment probably seemed in any one lifetime inevitable and unchanged. The ancient human acceptance and affirmation of a generous and gifting world was replaced by dreams of plenty in circumstances that made their fulfillment possible only in boom years. Domestication would create a catastrophic biology of nutritional deficiencies, alternating feast and famine, health and epidemic, peace and social conflict, all set in millennial rhythms of slowly collapsing ecosystems.

The complexity of social problems associated with domestication are difficult to understand but may have been due to sedentism. Was it because they quit being nomadic that primitive peoples became subject to scarcity and greed for things? There seems to be little doubt that political complexity increased with sedentism, but was that the result of power struggles over resources or the subtle effect of the proximity of one's neighbors, of being fenced in?[2] Perhaps the containment and the struggles for property and power cannot be disentangled. The potlatch people, sedentary fishermen, have the same troubles of power and influence that beset planters. Social conflict and competition arise in both cases, implying that sedentism is indeed at the heart of the problem.

Genetically the process of domestication is no different than adaptive change among wild species, a parallel which Charles Darwin intuitively recognized and which accounts for his interest in domestic pigeons and other farm animals. It takes only about fifty generations to alter a group of animals to the extent that it can be distinguished from its wild cousins. The production of new breeds and varieties of cats and dogs by humans demonstrates how rapidly "evolutionary" change can occur when directed by human selection.

The crucial factor in the keeping of animals that results in their biological alteration and renders them unfit to live in the wild is not simply captivity. Their genetic makeup is not altered by confinement. It is *breeding* in captivity that changes their genetic constitution. Selection of animals for visible "desirable" traits (size in dogs, milk in cows, wool in sheep) may make them unfit in other unseen ways (smaller brains, bone and skeletal problems, abnormal development, etc.). There is a self-culling in inbreeding, as some die or will not reproduce in captivity, but this does not offset undesirable traits that may be passed on.

Wild ecosystems have a higher diversity index (number of species per number of individuals), more niches, greater stability, higher net primary productivity (with less effect on the whole by the removal of a single species), higher structural and functional complexity, and greater population stability than cultivated systems.[3] The consequences for the captive and domesticated animals were reduction in size, piebald color, shorter faces with smaller and fewer teeth, diminished horns, weak muscle ridges, and less genetic variability.[4] Poor joint definition, late fusion of the limb bone epiphyses with the diaphyses, hair changes, greater fat accumulation, smaller brains, simplified behavior patterns, extended immaturity, and more pathology are a few of the defects of domestic animals. All of these changes have been documented in direct observations of the rat in the nineteenth century and by archaeological evidence and animal breeders in the twentieth century.[5]

The total number of species domesticated is minuscule compared to the number of wild forms. But weedy, wild forms, incidental parasites, and other plant, insect, arthropod, and rodent fellow travelers accompanied the domestic organisms and became interlocked with them as agriculture spread. An association of plants and animals emerged together with the human social and technological accoutrements of agriculture. As this human-dominated association replaced wild communities, drastic alterations were wrought in the microbial flora and invertebrates of the soil and water. So long as there were relicts of the wild habitats, the smaller, unobtrusive wild forms survived at the fringes or in the wild places between human settlements, while the larger mammals and birds tended to be excluded as competitors or were overhunted. But as people began to till the earth, other species were categorized as the enemy.

THE TRANSITION from the hunter/gatherer to the agrarian way of life took various paths, depending on circumstances, but in all cases it brought about similar changes in lifestyle and worldview. Although hunter/gatherers who had become more sedentary did begin some minor forms of cultivation, neither tribal histories nor the archaeology of hunting/gathering peoples shows that they readily embraced farming and herding. Instead, invaded for their land by both pastoralists and farmers, they were conquered. The small remaining subsistence or "Neolithic" gardeners or horticulturists today retain some of the old hunting/gathering ways, keeping traditions of sharing, the men seeing themselves as "hunters" even though slowly corrupted by a "warrior" concept. Many farmers and their city counterparts, having been conquered or displaced, are themselves refugees from better times. As though in some disastrous contract with the devil, they traded their social freedom for authoritarian regimes, the illusion of control in barren natural environments, and slavery in the garb of security.

The invasion of the homelands of hunter/gatherers is described in James Woodburn's description of the Haida, in Richard Borshay Lee and

Marjorie Shostak's work on the !Kung San, and in heavily documented descriptions of American Indian hunter/gatherers by their agricultural neighbors and Europeans.[6] These invasions have occurred over millennia against primal people with no warrior tradition, no idea of an organized army, and no psychology of defense. War and warriorhood probably grew out of the territorialism inherent in agriculture and its exclusionary attitudes and the necessity for expansion because of the decline of field fertility and the frictions and competitions of increased human density. With the rise of "archaic high civilizations," pathologies related to group stresses and the specter of scarcity in monocultures (grains, goats) meant, for all but a tiny elite, the loss of personal autonomy in the pyramiding power of conquest and the struggle for wealth.

When we compare the different economies of the past, we find the most striking features have to do with differences in the effects of ecology on the personality, especially compliance and obedience as distinct from self-reliance and independence. John Berry and Robert Annis studied differences in six northern Native American tribes and found "a broad ecological dimension running from agricultural and pastoral interactions with the environment through to hunting and gathering interactions." Corresponding psychological and practical differences were found between hunter/gatherers and planters who stored grains and roots or animal keepers with their tons of flesh on the hoof. Among the six tribes, agriculture was associated with high food accumulation, increased population density, and intensified social stratification. Hunter/gatherers were low food accumulators with a high sense of personal identity, social independence, emphasis on assertion and self-reliance, high self-control, and low social stratification. Berry and Annis see these composite differences in terms of "cognitive style," "affective style," and "perceptual style."[7] Robert Edgerton found distinct personality differences between farmers and pastoralists.[8] This difference includes perceptual habits and religious beliefs. (See Table 1.)

Although the aftermath of the agrarian way of life was filled with toil and scarcity, the earliest agriculture may have been a halcyon time for those who continued the older traditions of their hunter/gatherer ancestors and lived socially cohesive, small-group-centered lives. The process of

TABLE 1. FARMERS VS. PASTORALISTS

	Farmers	Pastoralists
orientation	soil cycle	sky power
sexual politics	matriarchal	patriarchal
animal symbol	snake	bird
commensal animal	ox	horse
funerary practice	burial	barrow graves
sacred terrain	springs, caves, etc.	mountain, sun
deities	polytheistic	polytheistic/monotheistic
chief deity	goddess	god
ultimate place	this world	the otherworld
source of help	local deities	messiah
pollution	organic and relative	puritanical
other life forms	subordinate	metonymic
conflict	defensive	expansive
mystery	earth's generation	the mind of god
authority	hatred of	respect for

transformation that propelled foragers into an agrarian way of life is worth revisiting. At that time of transition, soils were fertile and kept animals and plants still interbred with wild forms, adapting them to local environments and resistance to diseases—a resistance that the later, highly bred varieties and breeds would never have. This gene flow back and forth to the wild population can be seen today in the reindeer herds among Siberian, Eskimo, and Lapp peoples.[9]

In hoe or subsistence agriculture we usually find a greater diversity of plants, a polycultural system with small-scale mixed planting, and fewer specialized crops than with advanced agriculture. According to David Harris, the transition to the specialized systems from more general agriculture required not only genetic engineering but stratified societies.[10] This life, new to humankind, was a less interesting and challenging and more tedious way of life and required a "special mentality," according to paleontologist Wolfe Herre, "in order to accept the loss of freedom of a hunter's life."[11] This last statement may qualify as the great understatement of a century of prehistoricism.

Early in agricultural history, easygoing, subsistence societies made minimal introduction of domestic plants and animals, at the same time consciously resisting life in denser structures. Once villages were established, however, men began to fight over "the means of reproduction" and departed from the "modesty and conviviality" found in family-level societies. As populations of villages increased, geographical circumscription (expansion limited by mountains, desert, or sea) seemed to close around them, leaving nowhere to go, and more bullying, impulsive aggression, revenge, and territoriality took place. Scarcity of key resources and war became a threat to the daily lives of these horticulturists and animal raisers. As that economy changed (destroyed as a free and peaceful enterprise), the people found themselves in interdependent social groups where, because of the growth of the political economy they were forced into competition, warfare, and the necessity of defense against other groups.

In a stratified society people are divided into classes where individual freedom is limited. Wholeness and integration diminish because of the social effects of isolation and specialized roles. Of the Neolithic period that started about four thousand years after agricultural practices began, Wilhelm Dupre says: "The individual no longer stands as a whole vis-à-vis the life-community in the sense that the latter finds its realization through a total integration of the individual—as is the case by and large in gathering and hunting economies—but becomes part of the social structure in the function and role he selects or is selected for. It is a process in which natural alienation due to the psychological make-up of man assumes lasting forms."[12] The endogamous, secret, competitive nature of seed and root-keeping created the kind of Neolithic High Culture that preceded civilization—a fight and defense-mindedness that emphasized cultural differences. This attitude was promulgated by intensified tribal initiation that generated tribal ideology and converted clan members into warriors.[13]

✦

While the early Neolithic with its small-scale community and rich environments may have been among the best of times, most agriculture of the past five millennia has not been kindly. The theocratic agricultural states, from the earliest centralized forms in ancient Sumer, have enslaved rather than liberated. Even where the small scale seems to prevail, plenty and conviviality are not typical in Bronze Age, medieval, or modern peasant life with its drudgery, meanness, and suffering at the hands of upper classes; the vulnerabilities of the crop because of inbred weaknesses and resulting malnutrition of people make life even harder. The schizoid nature of agriculture can be seen in its paradoxical combination of tedium and relief in numerous violent festivals or carnivals.

Among archaic states that formed as villages joined forces, vassalage and standing armies make their appearance. Surrounded by people, the individual lost freedom, movement, and role. Johnson and Earle describe what happens as a nation matures into an agrarian state and has to deal with disagreements between households ensuing from distances in wealth and prestige. They often live so close to the margin of survival that they visibly lose weight in the months before harvest. As we approach the agrarian state, peasant subsistence provides a poor diet, undernourishment, extreme competition, and meager security as markets are controlled from the outside.[14] The record is one of endless rounds of population increase and dependence on "starchy staples."

The emergence of the controlling state was the result of social as well as technological change. Lynn White, the historian, has shown the association between the rise of agriculture and the "control" of nature. Plant monocultures were developed through the use of wind and water for turning wheels for complex irrigation projects and the forced labor needed to keep them going. White calls attention to the importance of the replacement of the ox by the horse in the eighth century and the profound change in efficiency and productivity created by the horseshoe and horse collar. As the horse made working the farther fields possible, a switch to "assarting," a three-step field system of clearing, draining, and diking, created landscapes that have characterized Europe for the past thousand years.[15]

Primitive planters—autonomous subsistence farmers or gardener-horticulturists—share much of the forager's reverence for the natural community and the satisfactions of light schedules, hands-on routines, and sensitivity to seasonal cycles. But with the advent of the ox and the plow, irrigation, engineering, the prison of seasonal rounds, and the horse as cavalry, it all changed. Autonomy vanished. The landlord became a warden, the yeoman a serf, a peon, a tool of the system. Women's work increased more than men's in both time and difficulty, and more of them went to live with their husband's people than was the practice in hunting/gathering societies in which about 65 percent of the residences of the newly married were with the wife's family.

✦

How do planters and animal keepers view their place on earth, in the universe, and in relationship to other creatures? What is the cosmic vision of the agrarian state? Joseph Campbell regards sacrifice as the planter's central rite where grain crops are the metaphor of the soul. The liturgical offering of fruit, grain, or the ritual slaughter of an animal or person is a symbolic means of participating in the great round, a rite of renewal or greasing the wheel. Our ties to the seeds of cultivated grain, Campbell claims, somehow imparted the idea of survival beyond death. Ceremonially, this idea is given expression by sacrifice.

Campbell has given us a deeply moving description of worldwide ceremonial practices among civilized and agrarian peoples, including those at the heart of Buddhism, Islam, and Christendom.[16] Yet, all the theatrical activity of his book amounts to page after page of bloody violence. It documents the underlying murderous and suicidal character that became common in these cultures. The immolation of the god is the central theme of Christianity, a self-sacrifice to redeem the believer's soul. Christianity's hostility to nature was celebrated in its asceticism, an orientation common to other religions as well.

What "is the game of the gods?" asks Octavio Paz, speaking of the Aztecs. Geographies are symbolic and landscape is historical, says Paz, and we turn them into geometric archetypes, such as the pyramid, "the metaphor of the world as a mountain." The "playing field" of the gods is the pyramid, "the religious-political archetype" with its platform sanctuary and symbolism of hierarchy. "They play with time," he says, "and their game is the creation and destruction of the worlds." It was a bloody game where prisoners were sacrificed, a game where a solar cult demanded that the gods be fed blood to keep the universe operating, just as the sun "daily is born, fights, dies, and is reborn."[17]

The "game" we humans play on earth may be regarded as finite or infinite, a contest in which the loser is destroyed or the winning is always temporary.[18] Is the "game" the animals hunted by foragers? Or is it, alternatively, a contest with rules and sides? Or is it both? The philosophy of the hunt tells us that games are infinite. Life goes on and nature provides the essential structure in a rule-regulated cosmos. The ecology and behavior of the "game" animals become the metaphorical model of human society, the rules of one's own biological being, and the world's working or playing. Winning and losing are transient phenomena—some small part of the whole. Opponents are essential. One loves one's enemies. To destroy them in any final sense is unthinkable.

Somehow that sense of perpetual play and the brotherhood of endless but leisurely opposition has faded with our primal ancestors, its place taken by the need for complete victory, a final solution. The authoritarian decree, reiterated again and again, has been the death of the others, the defeat of nature, of germs, of wolves. It is all the same, an obsession with total supremacy, as though the objective were to obliterate all defeated foes, all pests, all disease, all opponents, all the Others. To end the game.

Sacrifice does accommodate the "problem of death," as Campbell claims, but it does so merely by domesticating death. Sacrifice reverses the hunter/gather idea of gifting in which humans are guests in life who receive according to their due; in its stead it substitutes offerings as a kind of barter with blood as currency. Agriculture—domestic crops, for example—is characterized by glorious abundance or desperation. Harmony with the world is reckoned in terms of mastery over parasites and animal competitors by enlarging the scope of the simplification of ecosystems and, ceremonially, by sacrificial rites of negotiation with gods with human faces. Ostensibly a participation with the cosmos, the sacrificial ceremony is only a thinly disguised bribe.

In this "New Age" in search of messianic solutions to modern problems and the recovery of a lost world, we have uncritically embraced the shaman as visionary, medicine man, guru, ecologist, cosmologist, and wise man or woman and accepted the model of shamanistic thinking as ecological and nature-friendly. Spontaneous healers, usually women, have always accompanied humans. But the shaman is a latecomer—part of the agricultural fear of curses and evil spirits, the use of intoxicants, the spread of male social dominance, the exploitation of domestic animals (especially the horse) as human helpers, and the shift of sedentary peoples toward spectatorship rather than egalitarian participation.

Among foraging peoples, healers appeared spontaneously and did not necessarily hold other powers, sponsor séances, go on vision quests, do magic tricks, or wield political influence—all of which were true of the later shaman. Esther Jacobson, a scholar on Scytho-Siberian cultures, has shown how shamanism emerged as a late expression of what separates us from nature and marked the decline of the great cults of the bear and the mountain. The veneration of terrain features—lake, cliff, river, mountain, and cave—that attached people spiritually to place reflects "archaic traditions which go back before shamanism," which became a male-dominated political practice. Also lost were "contrasting relationships of bear/woman and bear/man" that carried "totemic inderstanding of tribal origins."[19]

The shift away from affirmation and participation in palingenesis—the round of life—to an attempt to control it can be seen in the deterioration of the ceremony of the slain bear as it was influenced by the outreaches of agrarian thought. In primal form the festival was an egalitarian, ad hoc, celebration of the wild kill as a symbolic acceptance of the gift of food. Modern tribal ceremonies of the bear cult have all but disappeared or have been altered, as in the Gilyak and Ainu of East Asia who kill a reared bear, scheduling the death of an animal under human control—surely not a hunt.[20] The ancient ceremony degenerated to a shaman-centered spectacle of the sacrifice of a captive bear, deflecting evil from the village.[21] The animal cannot be the focus of veneration and the object of sacrifice at the same time.

✦

THE TRANSFORMATION of the ecosystem of hunter/gatherers to the controlled monocultures of agrarian communities was accompanied not only by a change in cosmic view, but by the social and political zeitgeist as well. As agriculture became more complex, the importance of kin connections was subverted by politics, spiritual connections to the landscape were disrupted, and ecological relations to the land and animals were forgotten. The agrarian community of domestication reduced the life forms of interest to a few score species, mirroring the situation in which the farmers saw themselves dependent on deities with humanlike, often perverse, and unpredictable actions. Their cosmos was controlled by beings more or less like themselves, from local bureaucrats up through greedy princes to jealous gods. No wonder agrarian cultures preferred games of strategy and folktales (rather than the foragers' myths and games of chance) in which the petty tyrants, portrayed as animal burlesques of their various human persecutors, were outwitted by clever foxes like themselves.

In time farmwork became harder and more routine and organization more elaborate. "Headman" and "big man" politics emerged in which some men ruled and others obeyed. "Defensive needs" became a major concern, according to Allen Johnson and Timothy Earle, for offenses committed by the other big men and their followers, clans, and armies in the next village or state. As resources diminished and the land was denuded and eroded, there was a surge in extractive and storage technology, along with self-serving rhetoric and hierarchic ceremony by those in power. The rise of centralized authority—monarchies, clerical hierarchies, bureaucracies, trade networks, military units—was the heritage of agriculture, beginning with storable crops, distributional networks, bookkeeping, currency, territorial protection, and war.[22] Murray Bookchin describes this hunger for central authority as "a mania for domination that created mythic 'needs' and systems of control so harmful to the communities they were pledged to service that they and their legacy of waste, destruction and cruelty now threaten the very existence of society and its natural fundament. Indeed, the domination of nature was to have its roots in the domination of human by human."[23]

The transition from a relatively free, diverse, gentle subsistence to suppressed peasantry yoked to a metropolis is a matter of record. Today's urban gardeners and neo-subsistence people clearly long for genuine contact with the nonhuman world of nature, independence from the market, and the basic satisfaction of a livelihood gained by their own hands. But the side-effects of agriculture cursed the planter from the beginning. Faced with forced farming, Chief Washakie of the Shoshones said, "God damn a potato." Sooner or later you get just what the Irish got after they thought they had rediscovered Eden in a spud.

Domination follows competition. A Jeffersonian image of agrarian independence may have motivated Liberty Hyde Bailey to write his turn-of-the-century book, *The Holy Earth.* Bailey says: "Man now begins to measure himself against nature also, and he begins to see that herein shall lie his greatest conquests beyond himself. . . . The most virile and upstanding qualities can find expression in the conquest of the earth. In the contest with the planet every man may feel himself grow."[24] Then he patronizes the nature lovers. "I hope that we may always say 'The Forest Primeval.' I hope that some reaches of the sea may never be sailed, that some swamps may never be drained, that some mountain peaks may never be scaled, that some forests may never be harvested."[25] His solution to the destruction of the forest is the agrarian solution: walls, enclaves as parks and Indian reservations, a world separated into some tokens of nature and, outside the enclosures, the real, practical world of heroic engineering.

As for the fences between farmers, they betoken more rancor than peace. Since farmers cannot move away from unpleasant neighbors as easily as hunter/gatherers, they exercise conflicts in other ways—by laws and courts, for instance. They reluctantly submit to authority, more often settling into lifelong mutual hostility to both neighbor and authority. Robert Frost's poetic notion that "good fences make good neighbors" does not mean that fences increase the sum of mutual goodwill within the community, certainly not the inclination to share, but rather that walls inscribe with finality the power of ownership and exclusion, reducing trespass and potential friction.

The enclave confines or excludes things so that those on both sides can proceed with an insulated life. Its model is the walled garden: the wild things are kept out rather than in. Zoos, to keep wild things in, having been prisonlike, are beginning to become gardens with animals. But no such enclaves can maintain sufficient populations to sustain the genetic diversity of a wild population. The inmates do not belong to the actual ecosystems of the world biosphere but to a kind of halfway house just short of domestication. While enclaves may serve as emergency measures, they encourage us to believe in a world divided. It is the mentality of the dragon—the monster who sits under the mountain guarding his pile of gold and his virgins. Or it is the chessboard—take my square (if you can), capture my pawns, deal with my powers.

To moderns, agonizingly bereft of ceremonial life, the village beckons with an irresistible nostalgia. City people have always idealized the country. The Greek pastoral poets, Roman bucolic esthetes, and later European rustic artists fostered rural fantasies among educated urban dwellers. Its images of a happy yeomanry and peaceful countryside were therapeutic to the abrasions of city life. As art this agrarian impulse produced a kind of spurious, parklike ecology, the vegetable world as a better metaphor.

From the outside the life of the peasant or the villagers of hoe agriculture seems spangled with celebration. The calendar of folk festivals that are now part of the stock in trade of tourism, the color photographs of parades, dances, feast days, carnivals, and religious holidays, seem without end to readers of the geographical magazines who have themselves lost so much of community life and communal commitment. Primal foraging is widely looked upon as monotonous and dull. By contrast the numerous festivals in Third World villages and the heterogeneity of urban life seem infinitely lustrous and desirable.

Only lately has an esthetic developed around farming in the folk poetry and painting of the eighteenth century—the sentimental image of the straw-hatted, barefoot bumpkin wandering blissfully among his pigs or hoeing his radishes in the soft glow of the setting sun. We may ask whether there are not such stereotypes in the recent books of such advocates of the farm, the claim of stewardship, and the natural satisfactions of farm life. Farming seeks its benign figure not in a pictorial esthetic but in a social morality. Wendell Berry, for example, makes the garden and barnyard equivalent to a spiritual esthetics and relates it to monotheism and sexual monogamy, as though conjugal loyalty, husbandry, and a metaphysical principle were all one. And he is right. Archaic peoples, like most modern primitives, probably viewed the earth as mother and nature as feminine. But with the advent of agriculture the identification of woman with fertile land, as a fecund mother goddess, took on added significance. She became the symbol of productivity and access to the hidden powers of the earth, an image that has gone against women ever since.

Around the world as the subsistence base eroded, forests vanished, and water alternated from flood to drought, the attitude among both farmers and herders toward Mother Earth changed from worshipful to idealized. The ambivalent attitude toward earth, mother, and woman was translated into a fanaticism about virginity that made women pawns in games of power and under the control of men as the touchstone of honor and vengeance. It also reflected a deeper sense of alienation from the underlying maternal powers of the earth.[26]

For hunter/gatherers the living metaphor of cosmic power is other species; for farmers it is the mother; for pastoralists, the father. For urban peoples it has become the machine. But the mother and the machine have merged in the mind of commerce and the growth economy of the corporate world, a final degradation to which feminists have not been insensitive.[27]

Hunters and gatherers of both genders had lived in a somewhat jocular state of conflict in small-group societies—a sort of bandying of words in an egalitarian world. As we have seen, agriculture lent itself to imagining gods in the image of humankind who controlled humans as they controlled domesticated nature and as men controlled women. From the beginning, men have probably always suspected that women knew something that they did not. But it was not threatening until the phenomena of rot, disease, fungus, and spoilage came with crops, and, finding no simple explanation of this turn of bad luck, they pinned it on witching women. As larger human settlements became stratified, male and female tumbled into life as opponents. Their icons—as, for example, represented by Greek gods and goddesses—became competitive with each other for power.

The Earth Mother was probably more important and more durable in the root and perennial plant societies of Asia where water was of primary importance. E. O. James, who wrote *The Worship of the Sky-God*, tells of the road taken by the grain-growers tending annual plants: "In arid regions and in oases on the fringes of deserts where water was the most urgent need, particularly with the rise of agricultural civilization, it was to the celestial powers who controlled the elements that resort was made . . . for the life-giving rain or inundation. . . . In almost every ancient pantheon from the Neolithic onwards the figure of the sky god recurred, primarily concerned with the weather and the atmosphere. . . . He was regarded as all-powerful and all-seeing, and readily became a Weather-god."[28] Like bad money driving out the good in an economic system, patriarchal monotheism emptied the temples of the Goddess.

A marked change in attitudes toward death took place as the agrarian life took over. The Incarnation, the embodiment of God in the human form, is a central tenet of the "great world religions." Yet death has been constantly revised into the form of "everlasting life" in order to refine it and deny the corruption of the organic body that is part of the natural processes upon earth. The funerary preservation of body and mortuary architecture of stone are its final statement. The Hindus have exceeded the West in "a repudiation of the 'gross' material body."[29] Cremation renounces a sensuous and bodily world by a symbolic escape from the organic aspects of the cyclic flow of elements. "I will not be eaten!" is the last desperate cry.

Joseph Campbell says that "in the planting societies a new insight or solution was opened by the lesson of the plant world itself, which is linked somehow to the moon, which also dies and is resurrected and moreover influences, in some mysterious way still unknown, the lunar cycle of the womb."[30] The planters did obsess over fecundity, perpetual crops, and the pregnancy of their women. But according to the archaeologist Alexander Marshack, lunar periodicity had long been studied by hunters, who made calendrical marks with notations on carved bone and antler. Campbell theorizes that the ritually preserved bones of animals killed by hunters spring magically into a new animal of the same kind—"the undestroyed base from which the same individual that was there before becomes magically reconstructed." Among planters, by contrast, he says that the bone disintegrates and will then with the help of the group of farmers "germinate into something else. . . . The planter's view is based on a sense of group participation; the hunter's, on that sense of an immortal inhabitant within the individual which is announced in every mystical tradition. . . . The two have yielded radically contrary views of the destiny and righteousness of man on earth."[31] The planter's "group participation" implies subordination in a chiefdom and a transcendent or escapist spirit, while the private and unique soul of the hunter and that of his prey are seen to persist.

People who deal with natural death daily and directly do not deny and hide it; nor are they likely to become coarsened by it unless it becomes commercialized. The interdependence of life, however, is likely to be obscure to those who turn the killing of food animals over to specialists who practice in secret. Those who fear death become politically and socially conservative and less tolerant of other species, other creeds, and any deviation from their own mode of life.[32]

The structure of pagan ceremony was altered as the human community became sedentary and as agriculture coalesced political entities into ever larger domains. Among hunting/gathering groups or hamlet and village peoples, ceremony, dance, and other rituals, were based on myths and metaphors that signified gratitude toward the whole living world and participation with other species in the round of life. In "Big Chief" societies, however, the purpose of ceremonial dress, appurtenances, and performances affirmed levels of subordination and displayed the order of political rank. The symbolic center of ceremony shifted from an encounter with otherness in its many polytheistic forms to human social hierarchy and its humanized gods—one more step toward the mirror world of Narcissus in which humans replace an elegant comity of beings by despiritualizing nonhuman beings and using their skins and feathers as power symbols.

✦

THE CONSEQUENCES TO SOCIAL, emotional, and physical well-being as people are forced from hunting/gathering into agriculture are far-reaching. As omnivores, humans are characterized by a diet of "an enormous variety of foods," depending on seasons and availability as well as preferences. But "with the advent of the 'Agricultural Revolution' . . . there was less time for hunting and gathering, and the need to specialize on the cultivation of the most productive crops resulted in a simplification of the diet. . . . It is known that with reduced food choices, infant health is threatened because proper weaning becomes difficult and there is a consequent risk of malnutrition"—the main medical problem in the children of subsistence farmers in Africa, Asia, and South and Central America.[33] Says James V. Neel, "The advent of civilization dealt a blow to man's health from which he is only now recovering."[34]

Among the human diseases directly attributable to our sedentary lives in villages and cities are heart and vascular disorders, diabetes, stroke, emphysema, hypertension, and cirrhoses of the liver, which together cause 75 percent of the deaths in the industrialized nations. S. Boyd Eaton, an M.D. and professor of anthropology, and Marjorie Shostak, a respected researcher in anthropology and author of *Nisa: The Life and Words of a !Kung Woman*, comment: "The difference between our diet and that of our hunter-gatherer forebears may hold keys to many of our current health problems. . . . If there is a diet natural to our human makeup, one to which our genes are still best suited, this is it."[35]

Because of the overuse of salt, dairy products, sugar, and even crops like maize with its increased tooth wear and dental caries, our teeth are worse than those of primitive foragers, as are our bones, joints, and muscles. Many of our diseases we assume erroneously to be inevitable with age. We have more osteoporosis, lung diseases, and deafness than ever. Although the average height of all Americans increased several inches between the mid-1800s and the present, due to increased caloric and protein intake and improved health in childhood, "we have not quite reached the height of Cro-Magnon hunters and gatherers living 25,000 years ago."[36] There is a lower mean age at death, as well, and more anemia. Epidemic diarrhea is largely a marker of sedentary people everywhere.[37]

Domestic animals are the reservoir for many human parasites, especially viruses. During the past few thousand years they have endlessly generated mutant or recombinant forms that attack people with strains of encephalitis, measles, diphtheria; epidemics of highly infectious diseases known as plagues; and numerous multicellular parasites. Because of agricultural land use, malaria has become a major cause of human death. Archaelogical records show that the Neolithic was marked by "a decline in dietary quality" due to a lack of "availability of quality protein . . . and an increase in the consumption of starchy plant foods. Lowering of the protein-to-carbohydrate ratio "increases serotonin levels and induces a 'craving' for protein." This explains the "meat craving that is reported among so many hunting-horticultural peoples today."[38]

The fat in beef and pigs is notoriously bad for health because the intramuscular saturated fat (marbling in steaks), characteristic of grain-fed cattle, is "an artificial product of domestication" that is lacking in wild animals. Seal, whale, and walrus fat, widely eaten by foragers in the Arctic, is unsaturated. Polyunsaturated fat, linoleic acid, is not synthesized by the body and is essential to good health. It is found in vegetable fats, nuts, seeds, insects, amphibians, birds, snakes, and other reptiles. It is low in ruminants such as domestic beef.[39] Long-chain fatty acids, found in greater abundance in wild meat, are necessary for brain development. These come from structural rather than adipose fat. You can get them in meat from the butcher, but domestic cattle often lack access to an adequate variety of seeds and leaves to make an optimum proportion of structural fats.[40]

Neither domestic cereals nor milk from hoofed animals is "natural" food in an evolutionary or physiological sense. We are subject to epidemics of immune reaction, cholesterol susceptibility, and the dietary complications that arise from too much or too little milling of grains. The human difficulty digesting cow's milk is mainly because of the adult insufficiency of lactase, the digestive enzyme for milk, a deficiency that runs about 50 percent among blacks, 30 percent among whites, 70 percent among Chinese, and 24 percent among East Indians.[41]

Vegetarians disdain the arrogance of piggish meat-eaters and the health hazards of additive-laden meats, but the vegetarian alternatives subject them to a kind of nutritional brinkmanship. They must get eight of the twenty amino acids that their own bodies cannot make, all contained in meat in optimum amounts. None are stored, and the lack of one impairs the utilization of them all. The alternative plant sources are cereals and legumes, the first low in lysine, the second in methionine, so that people with little or no meat must get combinations of legumes and grain (lentils and rice, rice and beans, corn and beans) and must locate a substitute source for vitamin B-12.

"No exclusively vegetarian society has ever been discovered," says H. Leon Abrams. Of 383 different cultures, all eat animal proteins and fats, and "esteem them highly."[42] C. H. Brown observes that "small-scale agriculture supports population densities many times greater than those permitted by a hunting and gathering way of life. . . . However, a liability . . . is that crops are susceptible to periodic failure. On the other hand, the food supply of foragers consists of wild plants and animals that are naturally resistant to drought and disease, so that these organisms rarely, if ever, 'fail.'"[43]

Vegetarianism ignores human omnivory both in physiology and in food preferences. Food takes longer to pass through the gut in herbivores because of the slow digestion of cellulose-rich and fibrous foods. Like the gorillas, one of our Australopithecene cousins went the way of barrel-bellied herbivory. The small intestine is shorter in carnivores. In humans it is about half the length between gorillas and lions—the pure vegetarians and the pure carnivores—with digestive enzymes to match.

Except for a few leaf-eating colobine monkeys, the higher primates are nearly all omnivorous and have been so throughout their history. One-third of their diet is vertebrate meat, crustaceans, eggs, and a range of invertebrates. The dicotyledon eaters—omnivores originally of the forest, like us, who like salads too, have short intestine-to-body ratios. Savanna grazers depend largely on the vegetative path beginning with monocotyledons, which have more fibrous and lignified substances. Such plants are the principal forage of the ungulate grazers and browsers, which have a complex, symbiotic gut flora lacking in humans. Cereal grains are part of the monocot system. Had our hominid ancestors not played in the wider game of the hunt, adding meat rather than tough herbage, we might have wound up with bodies like gorillas, browsing placidly and almost continually. Vegetarianism, like creationism, simply reinvents human biology to suit an ideology.

Except for a tiny minority, people everywhere, including farmers, prefer to eat meat. Anthropologist Marvin Harris says: "Despite recent findings which link the over consumption of animal fats and cholesterol to degenerative diseases in affluent societies, animal foods are more critical for sound nutrition than plant foods."[44] Among most tribal peoples meat comprises less than 50 percent of the total diet; the bulk is made up of a wide variety of fruits, nuts, roots, and vegetables. But meat is always the "relish" that makes the meal worthwhile, and close attention is always paid to the way meat is butchered and shared. In virtually all small-scale societies, meat and hunting take precedence over plant food and gathering/growing. Perhaps there is an innate bodily wisdom about nutrition, but the immediate reason for the prestige of meat is because animals are believed to be sentient and spiritual beings like ourselves.[45]

Although traditional sharing of meat as well as gathered food is based on widely differing social criteria, such as lines of kinship or other obligations, not gender lines, some writers attach meat-eating to patriarchy. Though all sorts of arguments are presented to support this position, there is no evidence that "patriarchal" societies eat more meat than other societies, that soldiers eat any more meat than farmers, that hunters in hunter/gatherer societies (in which men usually do the hunting of large game), eat more meat, pound for pound of body weight, than the women. Women and men in all kinds of societies and circumstances prefer meat.

✦

SPECIALIZED FARMERS have always been basic adjuncts to large societies and hence, are linked by psychological as well as economic ties to urban dwellers. The agrarian mode was (and is) unstable. City anxieties about food are therefore independent of city control. "Sooner or later," observes Robert Allen, "increasing population and demands on land resources led to subdivision and fragmentation and relapse toward bare subsistence economy . . . checked by the reorganization of agriculture on an estate or feudal basis with the inevitable consequences of serfdom and slavery . . . which, unless placated with 'bread and circuses,' represented a continual menace to the ruling classes and the security of the state."[46]

The fantasy of agriculture as bucolic is the city person's fiction, who sees nothing of the resentments, the drudgery, or the intellectual vacuum. Perhaps it should be called "the wooden shoe delusion"—that cute object sold in gift stores which conjures up the clean little Dutch boy with his finger in the dike, beautiful fat cows in the background, while in reality the wooden shoe was the precursor to the rubber boot, worn by those who had to walk about in wet manure. Economists have their own pipe dream. Douglas C. North and Robert Paul Thomas see agriculture as man's "major breakthrough in his ascent from savagery to modern civilization" leading to individualized property rights and improved labor efficiency.[47] Like others they seem unable to get past the notion that maximized productivity is the ultimate good.

The historian's assumption that farming favored more security, longer life and greater productivity has been challenged by a student of foragers, Marek Zvelebil, who says that "when the reassessment [of postglacial hunting and gathering] is complete, foraging in postglacial forests will be considered a development parallel with agriculture and one that, for a time at least, was equally viable as a form of subsistence."[48] The rural countryside seems a wonderful escape both from nature and from the city. The first sentence in the preface to an anthology on domestication by Ucko and Dimbleby begins: "The domestication of plants and animals was one of the greatest steps forward taken by mankind." [49] After all, the idyll of the family farm, the Jeffersonian yeoman, the mental and spiritual relief of a rural existence is a heritage of civilization. It seems to have what hunting/gathering does not: retrievability. The agrarian life is only a generation or two away—indeed, only a few miles away in bits of countryside in Europe and America. After all, it may incorporate some hunting and gathering, as though creating the best of all possible worlds. Such a gardenlike, subsistence-oriented horticulture shades almost imperceptibly from a foraging life. At this boundary farming was probably once relatively benign, a satisfactory way of being human without the colossal destructiveness to which "modern" agriculture and its urban doppelgänger have led us.[50]

Even so, if there is a single complex of events responsible for the deterioration of human health and ecology, agricultural civilization is it. At its worst, agriculture is industrial and corporate, poisoning the whole planet with chemical compounds not found in nature. It has made plants and animals into what geneticist Helen Spurway calls "goofies," the deformed animals whose wild genetic homeostasis has been destroyed.[51]

Notes

1. Kent Flannery, "Origins and Ecological Effects of Early Domestication in Iran and the Near East," in Peter J. Ucko and G. W. Dimbleby, eds., *The Domestication and Exploitation of Plants and Animals* (Chicago: Aldine, 1969), pp. 73–100.

2. Susan Kent, "Cross-Cultural Perceptions of Farmers and Hunters and the Value of Meat," in Susan Kent, ed., *Farmers as Hunters* (New York: Cambridge University Press, 1989), p. 1–17.

3. David R. Harris, "Agricultural Systems, Ecosystems, and the Origin of Agriculture," in Peter J. Ucko and G. W. Dimbleby, eds. *The Domestication and Exploitation of Plants and Animals* (Chicago: Aldine, 1969), p. 4.

4. Inherited pairs of chromosomes that are different in their genetic make-up (heterozygous) provide the organism with a wider range of characteristics for responding to environmental circumstances and changes. Whereas this is more apt to be the case among wild populations of animals and normal human populations where breeding is not controlled, the chromosomal pairs in domestic animals that are inbred are more likely to have the same genetic constitution (homozygous) and thus have less genetic variability and ability to adapt to environmental changes.

5. R. J. Berry, "The Genetical Implications of Domestication in Animals," in Peter J. Ucko and G. W. Dimbleby, eds., *The Domestication and Exploitation of Plants and Animals* (Chicago: Aldine, 1969), pp. 207–217.

6. James Woodburn, "An Introduction to Hadza Ecology," in Richard B. Lee and Irven DeVore, eds., *Man the Hunter* (Chicago: Aldine, 1969), pp. 49–55; Richard Borshay Lee, *The !Kung San* (New York: Cambridge University Press, 1979); Marjorie Shostack, "A !Kung Woman's Memory of Childhood," in Richard B. Lee and Irven DeVore, eds., *Kalahari Hunters and Gatherers* (Cambridge, Mass.: Harvard University Press, 1976), pp. 246–278.

7. John W. Berry and Robert C. Annis, "Ecology, Culture, and Psychological Differentiation," *International Journal of Psychology* 9(3) (1974): 173–193.

8. Robert Edgerton, *The Individual in Cultural Adaptation* (Los Angeles: University of California Press, 1971), p. 275.

9. B. A. L. Cranston, "Animal Husbandry: The Evidence from Ethnography," in Peter J. Ucko and G. W. Dimbleby, eds., *The Domestication and Exploitation of Plants and Animals* (Chicago: Aldine, 1969), pp. 247–263.

10. Harris, "Agricultural Systems," p. 7.

11. Wolfe Herre and Manfred Röhrs, "Zoological Considerations on the Origins of Farming and Domestication," in Charles A. Reed, ed., *The Origins of Agriculture* (The Hague: Mouton, 1977), p. 7.

12. Wilhelm Dupre, ed., *Religion in Primitive Cultures* (The Hague: Mouton, 1975), pp. 327–328.

13. G. Rachel Levy, *Religious Conceptions of the Stone Age and Their Influence upon European Thought* (New York: Harper & Row, 1963), pp. 89–122.

14. Allen Johnson and Timothy Earle, *The Evolution of Human Societies from Foraging Group to Agrarian State* (Palo Alto: Stanford University Press, 1987), p. 277.
15. Lynn White, *Medieval Technology and Social Change* (New York: Oxford University Press, 1970).
16. Joseph Campbell, "The Way of the Seeded Earth, Part I," *Historical Atlas of World Mythology*, vol. 2 (New York: Harper & Row, 1988).
17. Octavio Paz, *The Other Mexico: Critique of the Pyramid* (New York: Grove Press, 1972), pp. 76–84.
18. James P. Carse, *Finite and Infinite Games: A Vision of Life as Play and Possibility* (New York: Ballantine, 1986).
19. Esther Jacobson, *The Deer Goddess of Ancient Siberia: A Study in the Ecology of Belief* (New York: Brill, 1993), p. 46.
20. Tim Ingold, *The Appropriation of Nature: Essays on Human Ecology and Human Relations* (Iowa City: University of Iowa Press, 1987), p. 260.
21. Paul Shepard, *The Sacred Paw: The Bear in Nature, Myth and Literature* (New York: Viking, 1985).
22. Johnson and Earle, *The Evolution of Human Societies*, pp. 157–159.
23. Murray Bookchin, *The Rise of Urbanization and the Decline of Citizenship* (San Francisco: Sierra Club Books, 1987), p. 24.
24. Liberty Hyde Bailey, *The Holy Earth* (New York: Scribner's, 1915), p. 83.
25. Ibid., p. 151.
26. Jane Schneider, "Of Vigilance and Virgins: Honor, Shame and Access to Resources in Mediterranean Societies," *Ethnology* 10 (1971): 1–24.
27. Paul Shepard, *Nature and Madness* (San Francisco: Sierra Club Books, 1982).
28. E. O. James, *The Worship of the Sky-God* (London: University of London Press, 1963), p. 20.
29. Maurice Bloch and Jonathan Parry, *Death and the Regeneration of Life* (Cambridge: Cambridge University Press, 1982), p. 37.
30. Joseph Campbell, *The Masks of God*, vol. 1 (New York: Viking, 1959), p. 180.
31. Ibid., p. 291.
32. Daniel Goleman, "Fear of Death Intensifies Moral Code, Scientists Find," *New York Times*, 5 December 1989, pp. C-1 and C-11.
33. J. R. K. Robson, ed., *Food, Ecology and Culture* (New York: Gordon & Breach, 1980), p. vii.
34. James V. Neel, "Lessons from a 'Primitive' People," *Science* 170(3960) (20 November 1970): 818.
35. Stanley Boyd Eaton and Marjorie Shostak, "Fat Tooth Blues," *Natural History* 95(6) (July 1986).
36. This evidence of our physical disability is taken directly from S. Boyd Eaton, Marjorie Shostak, and Melvin Konner, *The Paleolithic Prescription: A Pro-*

gram of Diet & Exercise and a Design for Living (New York: Harper & Row, 1988), p. 95. I am grateful for their statistics. Their book has every reason to hold forth with bold imagination, but it is a timid thing, unable to pursue its own logic, and unwilling to *say* that the life of our primitive forebears and contemporaries have and had a better life than we. It is the curse, I suppose, of scientists' reluctance to advocate anything.

37. J. W. Wood et. al. "The Osteological Paradox," *Current Anthropology* 33(4): 343–358.

38. Eric B. Ross, "An Overview of Trends in Dietary Variation from Hunter-Gatherer to Modern Capitalist Societies," in Marvin Harris and Eric B. Ross, eds., *Food and Evolution* (Philadelphia: Temple University Press, 1987), pp. 12–13.

39. Alan E. Mann, "Diet and Human Evolution," in Harding and Teleki, eds., *Omnivorous Primates*, pp. 22–23.

40. Robert Allen, "Food for Thought," *Ecologist* (January 1975): pp. 4–7.

41. According to geographer Frederick J. Simoons, malabsorption of lactose is due to an insufficiency of the digestive enzyme lactase that "hydrolyzes the lactose of milk into glucose and galactose, which can be readily absorbed." Although most animals experience a drop in the production of lactase after weaning, some individuals in human populations continue to be high lactase producers and lactose absorbers as adults. Simoons hypothesizes that high lactose absorption might have been a selective survival factor among populations experiencing inadequate amounts of other proteins and nutrients where milk from kept animals was readily available. These individuals "would be better nourished and healthier . . . and better able to protect and provide for" their families. See Frederick J. Simoons, "The Determinants of Dairying and Milk Use in the Old World: Ecological, Physiological, and Cultural," in J. R. K. Robson, ed., *Food, Ecology, and Culture: Readings in the Anthropology of Dietary Practices* (New York: Gordon & Breach, 1980), pp. 83–91.

42. H. Leon Abrams, Jr., "The Preference for Animal Proteins and Fat: A Cross-Cultural Survey," in Marvin Harris and Eric B. Ross, eds., *Food and Evolution* (Philadelphia: Temple University Press, 1987), pp. 207–224. See also Kent, *Farmers as Hunters*, p. 4, who says that "regardless of the subsistence strategy or mobility pattern meat and hunting are esteemed over plants and gathering."

43. Cecil H. H. Brown, "Mode of Subsistence and Folk Biological Taxonomy," *Current Anthropology* 26(1) (1985): 43–53.

44. Marvin Harris, *The Sacred Cow and the Abominable Pig* (New York: Simon & Schuster, 1987), p. 22.

45. Kent, *Farmers as Hunters*, p. 16.

46. Allen, "Food for Thought," pp. 4–7.

47. Douglas C. North and Robert Paul Thomas, "The First Economic Revolution," *Economic History Review* 30(2) (1977): 229.

48. Marek Zvelebil, "Postglacial Foraging in the Forests of Europe," *Scientific American* (May 1986): p. 104.

49. Ucko and Dimbleby, *The Domestication and Exploitation of Plants and Animals*, p. ix.

50. Among those who see in garden agriculture not only a worthwhile existence but a more practical solution to the difficulty of arranging our individual lives (rather than talking about hunting and gathering) are three American writer-farmers whom I admire enormously. They are Gary Snyder, the poet in the Sierra Nevada of California, who speaks so eloquently of the ties with the earth gained in place with the work of one's own hands; Wes Jackson, whose genius has flowered at his Land Institute in the prairies of central Kansas for shifting crops away from cultivation, from the use of chemical fertilizers and pesticides, and omitting overbred crop varieties; and Wendell Berry, the poet-farmer on his land in Kentucky, celebrating the best synthesis of nature and culture in the performance of such independence and virtues that subsistence fosters. I have repeatedly inveighed against all three for not pushing the thesis of an undiluted model of primal life to its conclusion. But of course I have known all along that there is no way, literally, for many people to achieve that final recovery of our truest being: to live wholly an authentic Pleistocene existence. And I know that simple farming with the protection of the immediate habitat is still possible for thousands of people—indeed, for millions, even in cities—if we can drive the corporate interests off the land. In the next-to-best of all possible worlds, I would welcome a triune of Berry, Jackson, and Snyder, empowered to take charge of the use of the continent, because I know that in spite of their grasses, legumes, or even potatoes that the wild world would survive in peace around them. May their Neolithic consciousness prosper—and prevail.

I have criticized them all, but I confess to a kind of in-house bickering. The quality of life that they themselves live, as nearly as one can see it from the outside, is superb. If the world could be put in their hands it would recover much of the best of the precivilized world of the Pleistocene. The bones I have to pick with them are surely those remaining from a shared hunt and meal—pieces to be mulled over (to mull, from a root word meaning "to grind" or "to pulverize"), which I take to mean that we are sitting at a fire together, breaking the femurs of deer to get at the marrow of things.

Snyder has said that the intent of American Indian spiritual practice is not cosmopolitan. "Its content perhaps is universal, but you must be a Hopi to follow the Hopi way." This is a dictum that all of us in the rag-tag tribe of the 'Wannabes' should remember. And he has said: "Otherworldly philosophies end up doing more damage to the planet (and human psyches) than the existential conditions they seek to transcend." But he also refers to Jainism and Buddhism as models, putting his hand into the cosmopolitan fire, for surely these are two of those great, placeless, portable, world religions whose ultimate concerns are not just universal but otherworldly. Yet from what I have seen of his personal life, there is no contradiction. I suspect that Snyder, like Berry and Jackson, is not so much following tradition as doing what Joseph Campbell called "creative mythology."

51. Helen Spurway, "The Causes of Domestication," *Journal of Genetics* 53(1955): 336–337.

VII
The Cowboy Alternative

In modern eyes cattle keeping is a rustic, sometimes colorful, and very marginal way of life. Yet pastoralism was one of the two great paths leading into the civilized world, and without its myths, traditions, and economy the modern world would be incomprehensible. The slow fusion of the earliest sedentary agriculture and the emergent ideology of the pastoralists between about six thousand and two thousand years ago gave us the first modern states. The long shadow thrown over the earth's ecology is that of a man on a horse, the domestic animal which, more than any other consolidated centralized power, energized the worldwide debacle of the skinning of the earth, the creation of modern war, and the ideological dissociation from the earthbound realm.

Imagine a progressive scenario in which early animal husbandry was simply a barnyard scene. As time passed, the ungulates destroyed forage nearby and keepers were forced to take their animals farther each day for grazing. The task passed from children to young men. Eventually this required days or weeks at a time when they were distant from the authority of the owners and elders. This was the cauldron of keeping and stealing, of daring and confrontation, making it possible for young men, increasingly proud and independent, to collect wealth in their own herds and escape the eyes of perpetual overseers. It put distance between them and the control of owners, fathers, and elders—least so where cattle keeping was mixed with farming among sedentary peoples living in rich and moist environments, more so where the environment was dry and the range damaged.

As herds became bigger and destroyed the pasture in the vicinity of the villages, especially in dry lands, cattle ranged farther and farther. At the same time, the demands of cultivation became increasingly restricting. Herding and farming became separate economies, and the two societies pursued their own needs and gods, seldom seeing eye to eye, although they remained mutually dependent as each produced things that the other needed.

The final separation of sedentary agrarian life and nomadic pastoralism emerged as the cattle keepers became mounted and were bound to a perpetual circle of movement with the seasons, returning to semipermanent settlements for part of each year. Cattle stealing was so central to the herders' way of life that it gave rise to a warrior class. The nomadic cattle people were enmeshed among themselves in a "segmentary principle," that is, groups and individuals alternately cooperated with others as needs arose for combining power to facilitate raiding, self-defense, and retaliation or competed with each other over water and grass. In this sense the horse intensified the need for collective solidarity, even though it undermined central control and magnified the means of dispute and aggressive individualism.

✦

THE COSMOLOGY of animal keepers was very different from that of farmers, whose gods and goddesses tended to be deeply connected to the soil and whose views tended toward a horizontal orientation as if determined by the way they looked out across the cultivated fields for the first signs of prosperity or disaster. Looking up to the spirits of storms, the sky, the wind, and the sun, the pastoralists developed a vertical, hierarchic pantheon, the most divine or most powerful deity at the top.

The earliest stockkeepers were peaceful, explains Esther Jacobson.[1] The belief in tree-trunk coffins and barrow and slab graves at first perpetuated the old Pleistocene foragers' three-layered concept of the cosmos (underworld, middle world, upper world) rather than signifying an escape from earth. Great Lady and Animal Mother images prevailed over male-centered prerogatives, and the cosmology emphasized clan lineage—symbolized by the foliate unity of the tree and branching antlers of the maral (deer)—and was ritually committed to the regeneration of life after death and the sacredness of stars, rivers, and mountains. All this was prior to centralized power among nomadic warrior brotherhoods.

In time the pastoralists, even more than farmers, looked increasingly for deliverance to the sky and the mountains, where storms were engendered, living as they did in marginal habitats, made more so by overgrazing by their cattle, where rain was scarce and bestowed by arbitrary gods. The grasses greened quickly after showers and were just as quickly eaten, forcing the livestock and their owners to move on.

To be on horseback is not only to be godlike but is also to see the earth itself as the underworld. It is no wonder that all "great" or "world" religions are embedded in pastoral motifs. Yet, caught in the regressive headiness of leaving the world, the mounted herdsman is puerile—in the sense that James Hillman discusses immaturity, in which up is better and therefore verticality in ourselves can be seen as a more immature orientation.[2] This orientation harks back to our semiterrestrial primate ancestors, with their own sense of the vertical, who survived by escaping from the ground into the trees. As former primate climbers we too may still be subject psychologically to instinctive arborphilia, in which safety and all good things are up, a vertical obsession that, millions of years down the evolutionary path, may have led to gnostic contempt for the earth.

Power over ever larger herds of ungulates, hence over people, swept all the earlier benign soft-herding away. That sense of power was further energized by the dream of flight triggered by the rapture of horseriding, the kinetic form of pyramidism, the architectural expression of leaving the earth, given the ecstatic realization of "flight" mythically taken on horseback. The horse was the end-of-the-world mount of Vishnu and Christ. As famine, death, and pestilence, it was the apocalyptic beast who carried Middle East sky worship and the sword to thousands of hapless tribal peoples and farmers from India to Central America.

All forms of escape from the earth—and the corollary of escape from the physical body—were probably unconsciously motivated by the desire to escape the degradation of the land, which began in Mesopotamia some eight thousand years ago. The sky dominated the world of the pastoralists and they seasonally beseeched a series of pantheons, headed by a sky god accompanied by a lesser earth goddess, for redeeming rain to ease the scorching aridity. The succession of spirits of the sky, older than the Jewish Bible, were presented as residents of an ethereal paradise who repudiated the earth as a true home. By the time of the first city-states in Mesopotamia, 2,000–3000 years B.C., the destruction of the land had generated chronic insecurity. To this, cattle keepers added the concept of moving to greener pastures, the ultimate power of the sky with its weather gods, the messianic rites of beseeching, and submission to distant powers,[3] attitudes that centuries later would produce our skyscraper mentality, that desire for transcendence or "ascensionism," the yearning, which can so dominate cosmology, to escape the earth.[4] Only the birds (who became angels) were freed from that to which humans seemed bound. Gordon Brotherston, a historian of pastoralism, notes "how thoroughly pastoralism has been inscribed in the twin ideological supports of Western culture: the Greco-Roman classics, and the Bible."[5] Pastoral thinking helped to define Western culture as it stitched together a larger ideology among animal keepers regardless of their geographically distant cultures.

✦

MOBILITY, LEISURE, SECURITY, even the opportunity to go on gathering from the wilderness, were present in pastorality. But the resemblance between peripatetic foragers and nomadic pastoralists is superficial. The warrior theme, the singular sky god, the myth of the wandering hero, the practice of sacrifice, the tilt of gender relationships, and the economy of domestic animals—all part of the pastoralist culture—were the antithesis of the hunters' world.

The hero came into Western Civilization with the Indo-European pastoralists, not only into Europe, but also into the cultures of the Greeks, Romans, Egyptians, Hebrews, and other Near Eastern peoples. The epithet for white races, "Caucasian," comes from the South Russian grasslands and Central Asian regions from which that horde of "Aryans" erupted southeast and southwest. Their ancestors had adored the wild horse—the onager—and cherished its milk and meat in domestication.[6] To the north they bordered on the Siberian tribes. Known to scholars from as far back as the fifth millennium B.C., their physical appearance, their primal form of our languages, and their proto-Occidental ways of life anticipate Europe. Most of us know them from bits of Scythian art and the fierce-looking horse folk who appear occasionally with their hunting eagles in travel magazines. They obtained (probably from the southern Near East) bovids and caprids, sheep, goats, and cattle, and in time created an economy and a cosmology that wedded their talents with the needs of the early agricultural states. Once these Asian pastoralists began riding their horses south they altered humankind's horizons forever.

The ancestral hunters' perception of the ecological symbiosis of predator and prey gave way to a new imagery of warriorhood. Esther Jacobson documents the shifting changes of the Scytho-Siberian nomads of the Crimea (north of the Black Sea) who hunted, fished, herded deer and other animals, and also grew grain. She describes the rich, peaceful culture in the seventh to the sixth millennium B.C. whose cosmogonic source of life and death was the Animal Mother in the form of a deer. As the culture changed through millennia from deer herding to horse and cow keeping and finally to mounted warriors (they were accomplished equestrians three thousand years ago), the imagery of the benign feminine figure of the Deer Mother was eclipsed by the parable of animal predation involving a feline and a horned antlered animal and later zoomorphic forms. The sacred deer of their hunter forebears was traded for the newly domesticated horse, which in the course of the changeover was ceremonially masked and adorned with false antlers. By the first millennium B.C. the deer had lost its mythic symbolism and was replaced by deer masks or antlered headdresses worn by horses and ornate saddles and tack. The horse had become the vehicle for telling the mythic story of predation. The Scytho-Siberian nomads retained much of their Neolithic religion until they encountered settled civilization and "aggressive statehood." They were the perfect candidates for what would be cavalry, their clannishness subordinated to military order, themselves seduced by the glory and booty of war. The Great Lady, the Deer Mother, the tree of life, and its feminine symbols were then subject to masculinist and warriorlike ideals in the hands of Bronze Age cultures with their "Hellenized preoccupations." As the old ways perished, Jacobson points out, "the archaic pantheon became focused on conventionalized and grim reflections of death. The theriology of their ancestors was reduced to a decorative animal style in art."[7]

Once people began to keep herds of animals such as reindeer and cattle they presumed their relationship with all animals to be a kind of herding. The Chukchi, for example, North Asian reindeer-keepers, saw the ownership of wild herds by *Picuucin*, the master, as corresponding to their ownership of domestic herds. This notion projects the tame herd idea on all of nature, so that wild things become pawn objects in transactions between true persons, that is, between humans and animal spirits.

Horse-herding peoples lived mostly on cattle and other ungulates, which were the commodity and currency of choice. Cattle were embedded in all social transactions, the measure of debt and wealth, the objects of affection and theft, the standard of esthetic ideals, the principals in the mythology, the major booty, and the main means of livelihood. The libations of milk and butter, the sacrifice of cattle and sometimes horses, the elaborate burials of aristocrats and their horses, all were central to their religious life.

The primal thought of Pleistocene peoples, had nothing to do with the concept of sacrifice, which Claude Lévi-Strauss sees as premised on a fundamental principle of substitution: "Please take this instead of the rest of us."[8] The idea of sacrifice was given definition and vigor by pastoralists. In making sacrifice, a sacred grassy area, the seat of a god, was strewn with meat offerings, gifts being accompanied by songs, in which a priest announced what the gift-giver wanted, in India and Iran usually "cattle and sons." Such negotiation could only have occurred with a god who was conceived as more or less like men—full of themselves and their power, trade minded, with the attitude of bargainers in a recalcitrant world, utterly different in spirit from the gifting world of the people of the bear, elk, and salmon. A liturgy of sacrifice, generally seen as the posture of a humble supplicant, revealed a despiritualized natural world reduced to materials to be bargained. As it was practiced by the Indo-Iranian descendants of the Indo-Europeans, sacrifice was perceived as "a presentation that establishes a relation of reciprocity, calling forth a countergift in return."[9] Meat that had been shared according to obligation and custom among hunter/gatherers became a kind of gift in the pastoral cultures in which there was constant maneuvering to obtain the favor of powerful lords.

Rites were stylized re-creations of underlying myths. The Indo-European ceremonies emulated the "first" sacrifice when the cosmos was created from the dismembered parts of a primordial god and the initial immolation of a king. Ceremonies preceding cattle stealing enacted another story involving a man, with the help of intoxicants and a deity, in conflict with a three-headed dragon—in its Hittite version *Illuyanka*—who represented the hunter/gatherer aboriginals obliterated by the Aryans.[10] A third ceremony and myth involved "the ideology of man as wolf," about a warrior who could become a wolf by donning its skin, "the highest accomplishment of the warrior's art, at once terrifying and glorious."[11] Wolfish rage was a "high art" for Indo-European men and their Persian descendants the Indo-Iranians.

Middle East culture, caught in the dichotomies of desert/cultivated, herder/farmer, and good/evil, ended Pleistocene thought. Traditionally the Indo-Europeans lived in tents in tightly clustered villages. Their genealogy was patrilineal, their ultimate sovereigns were celestial, and their deities were martial spirits. The nomads and villagers were bound to each other via tribal raids or commercial exchange. Leadership was based on power. Emphasis was on patriarchal endogamy, a double standard of sexual morality, and, in the Mediterranean, intense monotheism with demons, spirits and saints, veiled women, virginity, chastity, modesty, shame, kin-group loyalty, honor, hospitality, and an all-pervasive religiosity that made up this Western perspective.[12]

The hero, the warrior, and the cowboy are almost inextricable. For the most part of history they are all connected to horses or boats, although the Indo-European root looks especially to the horse. The energy of pastoralism had to do with the dynamic between freedom and authority. Farmers and tradesmen were tightly bound to village-centered control. Elders, whether men, women, priests, medicine men, bureaucrats, chiefs, mothers, fathers, or uncles, regulated the daily lives of the young and the ceremonies that marked the stages of life, marriage, communal obligations, and commerce. Young men particularly chafed under these circumstances. Opportunities for them to vent frustration were built into festivals, intervillage competition, races, and other liberating celebrations. But these were only cracks in the barriers of political authority and social rank and the village control of institutions with multiple, fearsome ghosts and gods.

Pastoral warriorhood was not a response to conflict among states in the modern sense. It grew, rather, from the complex of cattle stealing and the defense against it. Cattle raids and then "stealing back" were a kind of social dialectic among nomadic herdsmen everywhere, arising in response to the extensive patterns of animal grazing. Bruce Lincoln, author of *Priests, Warriors and Cattle*, says that warring "for cattle was a noble activity protected by the warrior god and sanctioned by myth. All the Indo-European peoples seem to have pursued it zealously with a sense of supreme confidence and self-righteousness."[13] The same was true of the non-Indo-European cattle peoples of East Africa and South America, particularly where they had steeds, and was so prevalent as to become part of the mythology of the anglo cowboy in the North American West and the gaucho in the South American Pampas.

✦

FROM THE POINT OF VIEW of herders, nature appeared to exert mastery in the way that they themselves exercised dominion over women and cattle. The bear, elk, and salmon, once the emissaries from the domain of the animal spirits to the human camp, were replaced by the shaman who made spiritual excursions from the camp into the sky world. The equilibrium of the golden age of the early Neolithic was shattered as bear mythology became extinct and the shaman became the adventuring hero. The holiness of the horse was infused with other and still older powers stripped from the wild. The elk (in America the moose) was among the most sacred animals of the late Pleistocene hunting cultures of Central Asia. Its demise as a holy animal of choice was followed by the advent of the newly domesticated horse, fitted with false antlers, described by

Jacobson, to signify that it had superseded the deer and elk deity, a "reminiscence" in the social transition from "deer herding" to horse and cattle. In its beautiful headdress the horse became "a vehicle for a complex of mythic ideas lodged in antlers, in horns, and in the theme of animal predation."[14] More than that, it incorporated the lost special divinity of birds and the sacred elk, known in the zodiac as Kheglen, and even, with the burial of horses in the tombs of princes, some of the powers of reincarnation originally derived from the figure of the holy bear.

The Indo-Europeans emerged from Central Asia as "a wave of shamanist horsemen," dispersed south in the third millennium B.C., and expanded along a broad front, sending contingents southeast into India and southwest into Iran and beyond, as far as Greece and North Africa.[15] Their values and ideas made up the sacred books, the Indian *Rig Veda* and the Iranian *Zend Avesta.* In the Indus Valley they destroyed the Harappan planter civilization of goddess worshipers and brought the male warrior, the Vedic belief system, and horses to the Indian pantheon that would eventually absorb them as part of Hinduism. In India the Code of Manu required that the king in his inauguration, enacting the myth of Prajapati, stand on a tiger skin, ritually mimic a cattle raid, and preside at a horse sacrifice. His consort, Sri-Laksmi, the goddess of royalty, sprang from Prajapati's mind much as Athena from the head of Zeus. The Indo-European invaders in 1500 B.C. brought horses, horse veneration, and the horse sacrifice, the epitome of all sacrificial rites, reserved for coronation ceremonies. The story of Vasistha and Visvamitra, "The Bovine's Lament," is about a stolen cow and a feud between warriors and priests, a divided jurisdiction in which the priests presided over the sacrificial rites and their myths, the herders or warriors over the stealing and recovery of cattle and associated myths.

The Indo-European shamanistic heritage is evident among the Greeks in the hero Perseus—the Greek betrayer of the feminine traditions of oracular and collective intuition from which he originally came, hypocritically wearing the shaman's gear, wallet, cap, sandals, and shield, "in the cause of descent from father to son; of politics, not religion; of rationality, not divination or possession."[16] The original visionary healing by individual women or men had been associated with the flight of birds who came to the healer. The professional shamanism that succeeded it was most fully developed by Indo-European pastoralists who reconceived the shaman's flight on the horse. The shaman, who had earlier departed the village by climbing a central tree or pole and taking flight, or who rode the drum, would instead leave by visionary horseback.

Riding a swift horse was the nearest experience of humans to intoxication and flight—to that ancient vertiginous excitement of the swaying tree. The mythical winged horse emerged as Pegasus sprang from the neck of the Medusa, decapitated by Perseus. As Medusa, the old goddess with her snakes, faded, her sacred horses were stolen by warrior heroes. "The wild and powerful thrust" of the "hoofs and beating wings" of Pegasus and the other winged horses of legend, says Butterworth, "imperatively demands the means and the knowledge" of control.[17]

WHEREVER THE INDO-EUROPEANS encountered indigenous cults, the victory over them was mythologized as a battle between a sky god and earth dragons such as the Greek Titans and Typhons. As it happened, the Indo-European incursions into the valleys of the Tigris and Euphrates corresponded with the zenith of great city-states such as Ur, Kish, and Lagash. These cities grew up from the rich monocultures of the riverine floodplains of the Near East. Their divine kings were seen in the sheep/goat idiom, an image of the benign pastorality in a mixed agriculture, as "the shepherd at the head of his flock," the defender against predatory enemies represented and then symbolized by the lion. Such autocracies had already begun the ideological move away from deified maternity, but not so far that the semidivine regents gave up the stories of being suckled at the breast of a goddess, or that priestesses did not still rule temples dedicated to one or another goddess of fertility.

The sacred nuptials central to ancient Mesopotamian agrarian renewal rites, seen as necessary to the success of the crops upon which a growing population depended, declined in mythic force just as the impact of Indo-European cattle-keeper invasions and three thousand years of soil loss and deterioration of vegetation, aggravated by climatic changes, made itself felt. The cutting of forests for construction and for fuel needed to heat, cook, make quicklime, and smelt metals, the destabilization of the water and soil by deforestation and overgrazing, and the salination of croplands may at first have intensified the worship of the sky god, Tamuz, son of the earth and water, who was dependent upon his mother/consort, the divine restorer of the seasons, to whom he looked for "release." Disorder in the basic ecosystems of the watershed and its life did not bode well for political or religious stability. The plant motif, embodied as seasonal renewal, with its emphasis on fertility in the earth, gradually lost ground to a heroic style shaped after the adventuring warrior and competitive pastoral society with its appeal to a distant sky god, instead of the village spirits, and the celebration of theft and recovery by countertheft, paradigmatic control over animals and women, and disdain for the earth. The sky god was imagined as a weather god, an outsider, a messiah who rides in to save all the people much as raiding parties of kinfolk or friends galloped in to rescue the stolen cattle or smite the enemy. Assistance in time of crisis or to augment a raid grew from pastoral society, but it became a major metaphor and mythic story by Hebrew times. The dream of the messianic savior became the Christian redeemer.

✦

ALTHOUGH THE FIRST GREAT CITIES were engaged in a network of exchange, the conflict and competition among them simmered, awaiting a means of military consolidation that transcended the limits of ragtag armies of erstwhile slaves and farmers on foot. As the nomadic pastoralists came down out of Central Asia with their horses, the war and turmoil that subsequently dominated five thousand years of Near Eastern history began with the transformation of horsemen into professional cavalry, hired by the chiefs of the great cities. The unique contract of horse warriors with the kings and priestly powers of the irrigation theocracies produced the divisions of the tripartite city-state—the ruling, the warrior, and the working classes—which we call civilization.

Nomadic pastoralism was by nature demographically dispersed: wild animals could hide in the cracks between tribes, and chiefs could salvage some of the traditions of the hunt. But when kings and priests formed an alliance in the ancient Middle East and hired armed and mounted herdsmen, they developed the capacity to reach out and subdue the opposition, natural and human. By joining with the state, this cavalry was able to redirect the conflicts to outsiders. Warriors—as mounted archers or on thundering chariots—periodically smashed and then occupied the cities of the ancient Near East and Asia Minor. The lust for power of rulers and the armed mobility of erstwhile pastoralists created civilization's war against the wild world and set the stage for the Near Eastern wars against each other that continue to this day.[18] The warrior in his most asocial form as the wandering hero is still recognizable as the déclassé cowboy, playing the half-mythical role of messiah and unable to submit to the tethers of civilized society.

The role of the horse and warrior in social control carried a deeper significance as a sense of power. Machines and cavalry made their appearance together. As Lewis Mumford puts it: "The instruments of mechanization five thousand years ago were already detached from other human functions and purposes than the constant increase of order. . . . 'Mass culture' and 'mass control' made their first appearance. . . . The ultimate products of the megamachine in Egypt were colossal tombs" and "as in every other expanding empire, the chief testimony to its technical efficiency was a waste of destroyed villages and cities, and poisoned soils: the prototypes of similar 'civilized' atrocities today."[19]

A parallel history occurred in precolonial South America. Of the four cameloids in South America, the Incas had domesticated two, the llama and the alpaca, by six thousand years ago. From present-day Colombia to Chile, says Gordon Brotherston, there were programs of "breeding, distinguishing, and counting types and ages of beasts down to the minutest detail."[20] Mounted on llamas, there was "a state army unparalleled in America" pursuing "policies of permanent territorial gain." All llamas were state property, granted as capital to settlers in conquered country. "Like their flocks, the subjects of Tahuantinsuyu could be considered contained and penned, pastured elements of the great *Pax incaica*, safe as such from the threat of enemies and the barbaric wild beyond its rim." Ceremonies equated the flock with the folk. Much as Psalm 23 declares, "The Lord is my shepherd," the Situa hymns highlight a monarchy endorsed by a pastoral creator principle: Viracocha, a supreme shepherd superimposed over local deities, and prayers comparing the self to the domesticated vicuna. (One is reminded of Jean-Jacques Rousseau's essay "On the Origins of Inequality" in which he compares human exploitation to that of animal herds and domestication to subjugation—which was perhaps more homology than analogy.) The Inca census was calculated with a quipu, a device made of brightly woven cords, used for both herds and people to record "performance." The flock economy shaped ideology: the symbolic significance of the llama involved individual llamas of special colors, a celestial llama, ceremonial scapegoating, sacrificial rites, and literary celebration in a poetry of metaphors of Andean pastoral conventions, in the high courts of Tenochtitlan and Cuzco.

The subjugation of independent villages or groups by centralized autocracy, either urban or arising from the hierarchies of nomadic pastoralism, was based on mounted couriers. Its apotheosis in communications networks—and militarism—is horse/llama/camel-powered or its mechanical equivalent. Ideologically, local powers were weakened and provincial deities and sacred places were subverted.

✦

PASTORALITY—generally a leisured life for herdsmen, counting ungulates and talking about cows—had a poet spokesman in Theocritus, from whom we got not only Virgil but Petrarch, a tradition of "classical" pastoral poetry, and a literature of pastoral ideology from Kahlil Gibran to Antoine de Saint-Exupéry. Grazing and browsing produced a kind of pseudo-savanna landscape that could be celebrated in all the arts, landscapes that to this day mask the destructiveness of grazing ecology. Saint-Exupéry's *Wisdom of the Sands*, with its sanctimonious, chauvinist immaturity, its fawning adoration of mother, and scorn for the fallen woman, reveals the schizoid mindedness to which pastorality lends itself.[21]

The dominant sky god and monotheism are essentially pastoral inventions, but the male power structure in pastoral ideology and society also idealized the feminine and thus made woman a lesser being, one that had to be possessed just as the herded and hoarded animals. Even in a sheep and goat idiom as transhumance—the seasonal movement of animals and herders to different grazing grounds—took place, women were more tightly controlled and less trusted and virginity became more of an issue. The oedipal fixations of boys and mothers, virginity as a social ideal, the "protection" of women by their removal from centers of power, the promulgation of the fiction of the hidden power of the harem—such elements are more than just a disguise of the patriarchate. They are part of a psychosis of the herd-keeping mentality that became part of the Western mentality.

✦

THE BIBLE RECOGNIZED farming as a curse and Abel, the shepherd, was the favored son of Adam and, indeed, of God. His brother Cain, the farmer, jealous, killed Abel. But why was Abel favored in the first place? Their names suggest a mythological basis in social status. "Cain" is from a Hebrew word meaning "creature," while Abel translates from Assyrian "son." Being firstborn, Cain was the first in line after his father, Adam, to suffer the sentence of earning his living by the toil and sweat that went with expulsion from the Garden of Eden. There was nothing noble or civilizing about drudging in the heavy soil and sluice ditches of the Euphrates valley. The Cain and Abel story took several thousand years to reach Hebrew ears. Shortly after the domestication of cattle, agriculture was rent by the division of which Cain and Abel are the mythical representatives, a division not simply in terms of economics, but a deep split of the mind.[22]

It does not follow that pastorality and planting are synonymous with patriarchy versus matriarchy. There is no symmetry. There were no "matriarchies" at the level of the state, although very small groups were often matrilineal and matrilocal. The feminine was essentially the focus of a cooperating plurality within any economy rather than a cult based on power. Insofar as the seeded earth was evoked in the maternal image of the mother as giver and taker of life, working the soil ameliorated the warrior spirit. All economies except foraging are basically capitalistic and have "heads," or chiefs, who are male. With horses, agrarian communities could remain loyal to the earth/goddess powers, subsumed and regulated by male authority.[23]

The relation between sedentary farmers and nomadic herdsmen can best be understood in the light of a series of societies ranging from completely sedentary farmers who keep some hoofed animals, to slightly modified groups who raise crops but whose animal husbandry extends greater distances from a home village, to others whose cattle keeping takes them away for whole seasons even though they are not truly nomadic, to still others of a more completely nomadic kind. This series forms a continuum from farming to nomadic pastoralism. The characteristics marking the two economies are likewise graded in this distribution. At one extreme are the farmers, still rooted in their belief in soil divinity and its female aspects, hoarding their seed varieties, isolated from outsiders, keyed to festivals that release them from the drudging routines; at the other are the nomadic pastoralists, their faith in celestial divinity and its social metaphors, their shifting alliances and the fluidity of obligations, celebrating warriorhood and its formalities.

The pastoralists gave us militant monotheism and the metaphors of the shepherd, the pastor, the pastorale, and otherworldly heavens of bucolic paradise. The shift toward pastoral monotheism drained sacredness from other forms of life and diminished the spirituality of lower beings, human or nonhuman.

✦

WEDDED TO SPECIALIZED AGRICULTURE and the bureaucracy of grain storage, mounted pastorality made the modern city possible and provided its ideology. The state's covenant with herdsmen-cum-warriors excluded the farmer's cosmos since the deracinated urban personality was closer in spirit to that of the pastoralists. In such large organizations as the state, the remoteness of the king was compensated for by his symbols. In contrast to sacred plants, animals, springs, and trees the king was not so much perceived as conceived—not immanent in place because his divine "place" was the sky, where the sun was his avatar. His presence was everywhere and his light and burning power showed his authority. The patriarchal loyalty embodied by fathers and elders, which governed village communities of mixed farming and herding, was transferred among nomadic pastoralists to a cosmic figure: a sky father. While they were in no position to wholly reject the feminine side of life, their aim was to control it. In royal societies nature had a more objective quality, was less numinous. Large herd size bestowed high status. Joseph Campbell associates the rise of state military and male rulers with pastoralism. He describes it as "extinction of the ego in the image of the god," replaced by mythic inflation or "the ego in the posture of the god." In the celestial mythology of the West the moon—which was formerly male, subordinate to the female sun, and periodically dying and being reborn—became subordinate to a sun that rules the heavens, the weather, and the pastures of earth via rain from the sky.

✦

THE PASTORAL NOMADS seem to have kept or recovered something otherwise lost with the old hunting cultures. The ease with which distance could temper disagreement among hunter/gatherers produced more open personalities. Daily life tended to be novel rather than repetitive. Leisure and security were more accessible; alliances could be shaped or broken at will. Independence, few possessions, opportunity to hunt and gather, life in tents free of sewers, crowds, taxes, and bureaucrats, have created an almost mythic figure in the Western imagination of the pastoral nomad. But pastorality created a bimodal existence: hospitality and generosity were balanced by a taciturn sense of honor, shame, and vengeance, a focused concentration on ownership that predisposed greed and an orientation that placed the self over others.

The separate directions of pastoral societies and foragers were profoundly different. Hunter/gatherers gained prestige by sharing meat, pastoralists by hoarding it to maximize their wealth and by collaborating to control labor.[24] Their preoccupation with domestic animals and political strife, tighter male dominance, and the elevation of religious patriarchy to a sky god produced disastrous ideology and ecology. Where farmers destroyed only arable land, the hooves and teeth of ranging ungulates destroyed wildlands, upper watersheds, and whole forests. When they joined the irrigation municipalities of Ur and Kish and a hundred other First Cities of the West as horse cavalry, the pastoralists contributed a sense of domination and mastery inured by centuries of rational animal slavery: castration, driving, rustling, and butchering.

The social conditions among male antagonists in pastorality required childhood training in which the stolid endurance of pain and autonomous aggression were rewarded. The present-day Fulani of Nigeria, for example, are nomadic cattle herders, attentive to dominance and subordination among the cattle and themselves. Bulls show threatening behavior and hence resist the Fulani control of the herd. Boys begin herding at age six, attending to calves, which they discipline with sticks, beating the reluctant animals. Courage is a Fulani ideal. The code requires that provocation be answered with physical attack. Cowards are beaten, fighting is common. A stick is used, the *kokora*, with which boys practice. The *sharo* is a contest during festivals, of boys aged fifteen to twenty-five, involving a challenge, an exchange of beating with the *kokora*, by taking turns. Lack of fortitude is humiliating. Out of this comes a fearless, aggressive, dominant personality. According to Dale Lott and Ben Hart, who have studied Fulani herdsmen, "if a herdsman has the sort of personality needed to display sufficient aggression to maintain his position as dominant over all the cattle in his herd, we might expect that his interactions with people would also involve assertive and aggressive behavior."[25]

✦

THE MODERN CONSUMER in the supermarket may have received his browsing instinct and tendency to perpetually take and move on from the equestrian drift of the pastoral mind across open country. Industrial agriculture and its urban conjugate—chain-store supermarkets and quick food vendors—have created an ambience of sameness, no matter where we go, and a universal placelessness.[26] Accessible as it is, the supermarket erodes something authentic toward which the genome points: an inherent need to actively engage in gathering, capturing, growing, and killing our food. A better world awaits us in the afterlife, we are told, a dream to which the old whip-waving, manure-treading herders clung, as they stared at the tail-ends of their "meat on the hoof." We no longer look toward a paradisiacal destiny with assurance; in fact, it is no longer needed, since it has been replaced by drugs and a paranormal world of virtual reality, a kind of narcoticized riff on the nature of place.

Judaism did not escape its part in this pastoral motif. The Jews filled the Old Testament with pastoral metaphors and perpetuated the Cain/Abel mismatch. Like the pastoralists, the Christian apostles were mobile and strident, trusting in the Good Shepherd. Through our religious heritage, we became the jumping-off people, interested in sky fathers and heavenly homes. Psychologically, the messiah complex is a cry for "mother" or "father." It means that our own carelessness or misdeeds will be salvaged by an outsider who arrives with the power to save us. For planters help comes up from the earth. For pastoralists it comes from over the horizon—the more etherealized the pastorality, the more likely assistance comes from above.[27] Mythically, pastoralism is a history of dramatic events, a chronicle of messianic interventions, apocalyptic overthrowings, crises and rescue, but the messiah syndrome grew from practical circumstances and became the source of divine intervention, modeled from the uncertainty of loyalty and betrayal. A posse that comes riding over the hill to save us in a time of crisis is its prototype. The messianic hope derives from the system of kinship and obligation that defines shifting support and antagonism, generalized in the Arab saying, "It is me against my brother. It is me and my brother against my cousin. It is me, my brother, and my cousin against the others." Belief and faith are more fundamental to pastoralists who must negotiate their salvation than they are to farmers who plant, pray, and wait.

Modern ranching further perverts old cattle-keeping practices that were noncommercial, noncommodity-oriented, and labor-intensive traditions in which men had personalized ties with their animals and were interested in taming them. They had individual knowledge of up to a thousand animals, their lineages, their names, their acquisition by debt, bridewealth, gift, stealing, or exchange, their peculiar characteristics, their ages, awareness of missing animals, herd behavior, reproductive history, and the social ties they each represented. These were people who ministered to and cared for the animals, and counted the herd as they went to sleep, and dreamed them in sleep.[28] Rather than commercial enterprise, their transactions included the paying of debts, bridewealth, gifts, exchange, and stealing cattle. Present-day pastoralists are on a parallel path with modern cattle ranching. Immersed in cow dung and urine, the Masai of Africa do not have a better ecology than the modern mechanized rancher; nor is the economics of Bedouin or Afghan nomads so much different from trading cows for cash. Educated modern ranchers have other devices for honing their cognitive skills than remembering the genealogy of the herd. Both modern pastoralists and ranchers emulate the mythic hero and celebrate male chauvinism, utilize extensive land on which their herds have replaced native ungulates, are extremely territorial, and romanticize their lives in song and story. The ritual by which modern ranchers celebrate themselves, analyzed in Elizabeth Lawrence's study of the rodeo, reveals the same underlying themes of the subordination of nature, immature ideals, the symbiosis of man and horse, territorial and spatial hegemony, the individual as macho hero, and misogyny.[29]

It is not my purpose to show farmers and pastoralists in a bad light against hunter/gatherers. The psychic abrasions due to their respective cultures, ecological desperation, and physiological destitution have raised personal levels of psychopathology, but individually they are no less devoted, loving, altruistic, religious, protective, honorable, and good than any other people. The potential for becoming as fully intelligent and mature as possible can be hindered and even mutilated by circumstances in which human congestion and ecological destitution limit the scope of experience. Life can offer substitutions for "nature," the environmental diversity needed for human fulfillment, but they are not infinite. Nor is one as good as another.

✦

PASTORALISM GAVE US domesticated animals and unleashed our desire for power. Horsepower began with animals hitched to grindstones and water pumps and became the standard measure for power.[30] Today power surges all around us perpetrating vast mechanizations and communications as well as ecological disasters and human ennui from which we seek escape. Pastoralism first gave us an escape mentality—the desire to get away from the things that enslave us, including those very powers we have sought—and has led us to one of the major themes of our time: consumerism as the great escape, the urge to buy everything from travel and automobiles to second homes, entertainment, and dope. In our journeys, from the beginnings of the West to space exploration, we have followed this power mania as though we were all infected from the beginning with the young male pastoralist's frantic desire to be free of elders, to find his place, to dream of overnight riches, to get his feet off the ground, to be transported.[31]

The enslavement of cattle, horses, llamas, camels, dogs, cats, and other domesticated forms is reality concealed under a crust of sentimental nurturance. Speaking of biblical shepherds, for example, Calvin W. Schwabe, a veterinarian-professor, has this to say: "Sheep-culture peoples were probably the first to possess behavioral qualities we consider *humane* as distinguished from bestial. From the pastoral occupation of these roamers under the stars emerged the humanizing qualities of gentleness, caring, compassion, responsibility, nonviolence, and contemplation, and these values were consistent with development of an interest in healing and beginning acquisition of the manual skills of a healing art."[32] The word "designing" is used to describe the Hebrew attitude toward nature, and it applies to all those lamb-carrying figures in pastoral imagery. The lamb will not only lose its wool and possibly its milk but it will eventually have its throat cut. If this is the ultimate model and origin of human caring and compassion, then we all have reason to suspect our pastors, brother-keepers, and parents.

Those who venerate farming would have us believe that the tender care of potatoes by farmers is for the potato's benefit. As for cattle, anybody who has been around them on the one hand, and around elk, deer, or any other wild ungulates on the other, knows what total potato-heads cattle are. But history's lack of tolerance does not speak for prehistory, in which the perception of animals was antithetic to what came later.

The transformation of animals through domestication was the first step in remaking them into subordinate images of ourselves—altering them to fit human modes and purposes. Our perception of not only ourselves but also the whole of animal life was subverted, for we mistook the purpose of those few domesticates as the purpose of all. Plants never had for us the same heightened symbolic representation of purpose itself. Once we had turned animals into the means of power among ourselves and over the rest of nature, their uses made possible the economy of husbandry that would, with the addition of an agrarian impulse, produce those motives and designs on the earth contrary to respecting it. Animals would become "The Others." Purposes of their own were not allowable, not even comprehensible. Our relationship to the nonhuman life on earth, a relationship lost with the Pleistocene, is no different than cherishing relationships that we are capable of developing with humans who are very different from ourselves. Helene Cixous and Catherine Clement suggest a course for humans: "Each would take the risk of other, of difference, without feeling threatened by the existence of an otherness, rather, delighting to increase through the unknown what is there to discover, to respect, to favor, to cherish."[33]

Notes

1. Esther Jacobson, *The Deer Goddess of Ancient Siberia: A Study in the Ecology of Belief* (New York: Brill, 1993) p. 58.
2. James Hillman, "Senex and Puer," in James Hillman, ed., *Puer Papers* (Dallas: Spring Publications, 1979), pp. 3–53, quoted in David Lavery, *Late for the Sky: The Mentality of the Space Age* (Carbondale: Southern Illinois University Press, 1992), pp. 43–44.
3. E. O. James, "The Worship and Nature of the Sky-God in Semitic and Indo-European Religion," in *The Worship of the Sky-God* (London: University of London Press, 1963).
4. Lavery, *Late for the Sky*, p. 29.
5. Gordon Brotherston, "Andean Pastoralism and Inca Ideology," in Juliet Clutton-Brock, ed., *The Walking Larder* (London: Unwin Hyman, 1990), p. 240.
6. Whether the horse was domesticated from *Equus przewalskii*, Przewalski's horse, or *Equus hemionus*, the onager, is a matter of disagreement among the experts. Wolfe Herre and Manfred Röhrs in "Zoological Considerations on the Origins of Farming and Domestication," in Charles A. Reed, ed., *The Origins of Agriculture* (Chicago: Aldine, 1977), p. 69, claim it could not have been the onager.
7. Jacobson, *The Deer Goddess of Ancient Siberia*, pp. 28–47.
8. Claude Lévi-Strauss, *The Savage Mind* (Chicago: University of Chicago Press, 1966), pp. 223–226.
9. Bruce Lincoln, *Priests, Warrior and Cattle, a Study of the Ecology of Religion* (Berkeley: University of California Press, 1981), p. 68.
10. C. A. S. Butterworth, *Some Traces of the Pre-Olympian World in Greek Literature and Myth* (Berlin: De Gruyter, 1966), p. 155, observes that shamanism and habitual intoxication are associated.
11. Lincoln, *Priests, Warriors and Cattle*, p. 126.
12. Emanuel Marx, "The Ecology and Politics of Nomadic Pastoralists in the Middle East," in Wolfgang Weissleder, ed., *The Nomadic Alternative* (The Hague: Mouton, 1978), pp. 41–74.
13. Lincoln, *Priests, Warriors and Cattle*, p. 131.
14. Jacobson, *The Deer Goddess of Ancient Siberia*, p. 36.
15. Butterworth, *Traces of the Pre-Olympian World*, p. 153.

16. Ibid., p. 167.

17. Ibid., p. 170.

18. This struggle and yet the inevitable upward destiny of humankind are given a full-blown treatment in Jacob Bronowski's book, *The Ascent of Man* (Boston: Little, Brown, 1973), which documents a kind of self-hypnosis in the course of history. The film that accompanies it shows Bronowski serenely riding across the desert and plains with pastoral Near Eastern Kurds whom he admires for their attunement to nature and close social community. It is as though he is caught up in a thinking man's Disneyland. The environmental cost of such tranquility is invisible, just as it is for the public with respect to the cost of all Disneylands.

19. Lewis Mumford, "The First Megamachine," *Diogenes*, Fall 1966, pp. 1–5.

20. Brotherston, "Andean Pastoralism and Inca Ideology," pp. 244–248.

21. Antoine de Saint-Exupéry, *The Wisdom of the Sands* (New York: Harcourt Brace, 1950).

22. Lincoln, *Priests, Warriors and Cattle.*

23. Ibid., pp. 195 and 206.

24. David Riches, "Hunting, Herding and Potlaching: Towards a Sociological Account of Prestige," *Man*19(2) (1984): 234–251.

25. Dale F. Lott and Ben L. Hart, "Aggressive Domination of Cattle by Fulani Herdsmen and Its Relation to Aggression in Fulani Culture and Personality," *Ethos* 5(2) (1977): 177.

26. One might wonder how modern society could combine the nomadic life associated with the pastoral aegis and the sedentism of the farmer. It has been accomplished through pseudomovement—that is, tourism, which may be defined as the affirmation of locus in the guise of travel, or, put another way, a form of travel in which a special industry creates the illusion of peregrination.

27. David Crownfield, "The Curse of Abel: An Essay in Biblical Ecology," *North American Review* 258(2) (Summer 1973): 58–63.

28. John G. Galaty, "Cattle and Cognition: Aspects of Masai Practical Reasoning," in Juliet Clutton-Brock, ed., *The Walking Larder* (London: Unwin Hyman, 1990), pp. 215–230.

29. Elizabeth Lawrence, *Hoofbeats and Society* (Bloomington: University of Indiana Press, 1985).

30. Lynn White, *Medieval Technology and Social Change* (New York: Oxford University Press, 1962).

31. Lavery, *Late for the Sky.*

32. Calvin W. Schwabe, "The Most Intense Man–Animal Bond," unpublished manuscript, n.d.

33. Helene Cixous and Catherine Clement, *The Newly Born Woman* (Minneapolis: University of Minnesota Press, 1986), p. 78.

VIII
Wildness and Wilderness

WILDNESS IS A GENETIC STATE. Wilderness is a place we have dedicated to that wildness, both in ourselves and in other species. The home of our wildness is both etymologically and biologically wilderness. Although we define ourselves in terms of nationality, race, profession, and so on, it is evident that the context of our being in the past is wilderness—to which, one might say, our genes look expectantly for those circumstances that are their optimal ambience, a genetic expectation of our genome that is unfulfilled in the world we have created. It is as though we need a shield to keep from being mashed by the great juggernaut of modern times. Imagine the weight of ten thousand years of farms and civilization rolling heavily along.

The mapping of genes on chromosomes is reported in almost daily bulletins from medical research specifying the location and identity of deleterious genes. Some anticipate that invasive techniques will enable us to replace bad genes with desired alternatives early in the life of an individual. Although pursuit of total health and perfect crops drives this research, at a less conspicuous level this research is also defining the genetic basis and reality of the "normal" or optimal human individual. What was often thought to be "cultural" or "learned" is now shown to have a genetic basis and experience or "education" is mostly a kind of facilitation. At last we approach the hard truth: being human is heritable. (Just as being chimpanzee is heritable: witness forty years of failed attempts to make chimpanzees human by rearing them in human homes.) In the light of genetic research, the tall tales told by humanists and sociologists for so long of children who, lost in the woods and suckled by wolves, grew into "wolf children" are clearly a fantasy. No matter what the circumstances, children will be children, not wolves.

The idea of wilderness, both as a realm of purification outside civilization and as a special place with beneficial qualities, has strong antecedents in the High Culture of the Western world. The ideas that wilderness offers us solace, naturalness, nearness to a kind of literary, spiritual esthetic, or to unspecified metaphysical forces, escape from urban stench, access to ruminative solitude, and locus of test, trial, and special visions—all these extend prior traditions. True, wilderness is something we escape to, a departure into a kind of therapeutic land or sea, a release from our crowded and overbuilt environment, healing to those who sense the presence of the disease of tameness. We think of wilderness as a place, a vast uninhabited home of wild things. It is also another kind of place. It is that genetic aspect of ourselves that spatially occupies every body and every cell. To "go into" that wilderness is something we do constantly. We are immersed in it. Our consciousness and our culture buzz around it like tiny lights, not illuminating a great darkness but drawing energy that makes a self possible.

Whether humans are "domesticated" has been argued for decades, but it is mostly a semantic problem. "Domestic" means a "breed" or "variety" created by the deliberate manipulation of a plant or animal population's reproduction by humans with a conscious objective. We ourselves are genetically wild rather than domesticated. The metaphors of "domestic" have confused this truth.[1] We are also tame, for almost any animal can be conditioned to accept the human environment, domestic or not. The tameness of captive wild animals is like our own tameness: it is conditioned to appropriate behavior in the household. Our tameness not our domestication, makes us at home in domesticated landscapes, in the sedentary life surrounded by household artifacts and the romance of the hearth and homestead, restraints in the interest of civic order that cloud the definition of "domestic." None of these affects the human genome and therefore its wildness.

The domestication of plants and animals typically produces rapid genetic changes and exaggerates both selected and unintended traits. Some deleterious characteristics carried by one gene are normally hidden in the mixed genetic bag, which is rather like a safety net. For example, if one gene carries a recessive trait that would make the individual susceptible to a certain disease, the person is protected from the disease by the other dominant gene. The anomalous features for which we select plants and animals at the expense of overall adaptability—often by breeding for pairs of recessive genes—are the heart and soul of domestication. Typically such forms cannot survive without human protection, in gardens, farmyards, households, laboratories, or greenhouses, because their overall stamina or intelligence has been sacrificed for special features.

Today we live not only in the structures we build but in domestic associations with dandelions, bluegrass, grainfields, and early plant successional forms that are often weeds—all together composing relatively stable-appearing, made landscapes whose durability is illusory, not so much in equilibrium as in a kind of ecological suppression caused by human and domestic animal pressures. One of the oddest and most compensatory things about this domestic landscape complex is the continued existence of genetically wild forms in it as microorganisms and, at its margins, wild legumes and flowering plants, insects, foxes, crows, langurs, shrews, and ourselves. None is bound strictly to the mosaic of domestic plants and animals in order to survive in the way that domestic life forms are.

Such landscapes induce in us an unease—a fugitive sense that we persistently misdiagnose because the symptoms tend to be social as well as ecological. The radical implication is that we, like the other wild inhabitants, may actually be less healthy in the domesticated environments than in those wild landscapes to which our DNA remains tuned. When we do find rural landscapes beautiful, it is probably because they superficially resemble the savannas of our evolution. Indeed, the patterns of typical agrarian life seem to echo on a rough scale the dim memory of mixed open and wooded land.

But that appearance is superficial. Our domesticated surroundings are human inventions—the results of empirical technology over the past ten millennia and scientific technology during the past three hundred years. Immersed in landscapes dominated by built and domestic forms, we are not yet confined to them, and our human potential is less in such artificial landscapes than in those places and cultures shaped more directly by the terms of our evolutionary genesis, where we are realized as mature individuals and communities of generous and peaceful character.

Like raccoons and bears, we are omnivorous, edge-dwelling forms whose movement through different habitats tends to mask our ecological constraints. Modern life conceals our inherent need for diverse, wild, natural communities, but it does not alter that need. Evidence for this deprivation is so omnipresent that we cannot see it directly, since much of it is expressed as psychic stress and social disorder. Masking the effects of deviating from the world to which we are adapted is the universal act of modern denial.

That we (like opossums and cockroaches) can endure deficient environments has been interpreted as evidence of human transcendence over biological specialization, a widely ridiculed error of dinosaurs and other extinct forms as a hateful trap. Some observers have insisted for generations that ours is a "generalized" species, while all around us the fossil animals made the "mistake" of becoming "too specialized." That the dinosaurs lasted 170 million years (indeed, still exist as birds) is irrelevant to those celebrants of the great reptiles' ephemerality .

Our self-deception culminates in the notion that the human brain, the magic means of our intelligence and mastery, is the instrument of our exception from the biology that has burdened and exterminated so many other species. Yet this is the same brain and the same nervous system now found in dysfunctional humans in the scarified lands of the Near East and Middle East, much of Africa, Asia, and all urban existence. What was a good (and very highly specialized) brain for positioning a terrestrial primate in a Pleistocene niche is evidently maladapted for life in the throes of its own glut of people and barrenness of nature.

We are not the generalized species we claim. Indeed, human ontogeny—the intricately structured, human life cycle—is, like our central nervous system, a delicately equilibrated biological complex. The paradox of what we have thought was unlimited adaptability and extreme specialization of the human will probably untangle its own contradictions in the twenty-first century. Then, perhaps, when we have taken our adaptability to the brink of physical and psychological endurance, we will discover that cultural choices, unlike our bodies, do not have built-in limitations and requirements. Constraints are not welcome in an ideology of unlimited expectations among affluent societies, where, in the rush of individuals creating themselves, the self is left as an open sore. Our cultural choices are rewarded or punished according to our given natures. Such constraints are part of a universal biological heritage, honed to a Pleistocene reality, to those three million years that ended about ten thousand years ago.

✦

IN THE LATE TWENTIETH CENTURY a renewed sense of human nature began to emerge. People began to question the assumption of inevitable progress and the premise of human dominion over, or exemption from, the "laws of nature." The shift away from this illusion—that we can be anything, go anywhere in space, or remake the planet according to taste—was foreshadowed by the words and works of a few hardy thinkers,[2] all of whom shared the best idea in the last five thousand years: Charles Darwin's concept of the evolution of life.[3]

Until recently we have portrayed human wildness in one of two contrasting fictions: the Noble Savage, living in or having lost his golden age of human perfection; or the Cave Man, a slavering brute, lurking at the fringes of humanity itself, destined to consort with the beasts as one of them. As I have discussed in Chapter V, both images are fictions of ourselves: the first as the lonely outcast of lost paradise, the second as a savage barely emerged from a hairy, grunting animality. For the Greeks, Romans, and Christians, the Wild Man was the product of the wilderness, deficient in morality and every other human virtue, and remains the not-yet-human of the past, above whom Progress and High Culture elevate us.

In either case, our "animal" state in the popular mind corresponds to what we wrongly deduced from watching the demented and stupid beasts of the barnyard. The only hope to escape such gluttony, lust, and violence was through moral rigor, religious salvation, or some kind of social amelioration that would block such destructive impulses. Despite its brilliant insight, Sigmund Freud's psychology of instinct was limited to a combative or sexual urge to be suppressed and controlled by rational thought. In such ugly visions of wildness and the nature of the self to which we are heirs, it is not surprising that the modern idea of wilderness's value has little to do with the biological ground of our being. It is predicated instead on esthetics, on a rational ethic of biodiversity, on the concept of a protective enclave for wildlife, or as "recreation."

There is, however, a new paradigm of primordial recovery. It models optimum qualities of human life not only in terms of philosophy and culture but also in food, exercise, and society, as these existed among late Pleistocene humanity and still exist in relict hunter/gatherer peoples. Wilderness, we now see, is not only an adjunct to the affluent traveler or "inspiration" for an educated class; it is also the social and ecological mold of our species, which continues to be fundamental to us.

Biologist Hugh Iltis writes, "Man's love for natural colours, patterns, and harmonies, his preference for forest-grassland ecotones which he recreates wherever he settles, even in drastically different landscapes, must be the result (at least to a very large degree) of Darwinian natural selection through eons of mammalian and anthropoid evolutionary time. . . . Our eyes and ears, noses, brains, and bodies have all been shaped by nature. Would it not then be incredible indeed, if savannas and forest groves, flowers and animals, the multiplicity of environmental components to which our bodies were originally shaped, were not, at the very least, still important to us? Would not such a concept of 'nature' be a major part of what might be called *a basic optimum human environment?*"[4]

Our general acceptance of the genome as a controlling part of our lives is changing, not because society has become more evolutionary in understanding, but through medical research. Our hereditary integrity is a reflection of a deep past that continues in us. Our health in the broadest sense depends on it. As we begin to see organic dysfunction and disease as the misfitting of our genome to contemporary environments that we have created, we move away from the notion of war against natural process and against wildness. We are Pleistocene hominids keyed with infinite exactitude to small-group, omnivorous life in forest/plains edges of the wilderness. An increasing number among us have immunological intolerance to milk and cereals and vascular systems clogged with domestic fats and cholesterols. We face decrepitude of body and spirit caused by sedentism, the psychoses of overdense populations, failed ontogenies, and cosmologies that yield havoc because they demand control over, rather than compliance with, the wild world—cosmologies based on the centralized model of the barnyard.

We did not recently begin to move toward better diet and exercise because of our sense of identity with life in the lost world of the ice ages, but because of our own symptoms of alienation from it. Most of us remain unaware that the remote world of the "ice ages" is where the criteria were established that determine whether our medical therapies are successful and whether we truly understand what recovery means.

✦

THE TIME HAS COME TO dispose of the notion of wilderness as the last zoo, as an exalted, beautiful picture, as a precious, exotic landscape, or as a storehouse of tomorrow's resources. Some wild things require vast spaces: our ancestors occupied home ranges of hundreds of square miles. Wildness, not wilderness, is the state against which we assess the weaknesses as well as virtues of civilization and its correlates—mass society, the use of fossil fuels, growth-oriented economies, monocrop agricultures, and the technologies of control resulting in dysfunction and pseudomastery that conceal our limitations with glut, comfort, and entertainment in a world of virtual reality. Although we seek therapy in the wilderness as though it were relief from real life, the effort to recreate, to study or appreciate the balm of wilderness, to compose a journal of self-discovery are, culturally speaking, merely palliative. Public concern over the increasing rate of extinctions and the worldwide diminishing of biodiversity is, in the end, not altruism, ethics, or charity, nor has it to do with the paintings of John

Martin and Alfred Bierstadt, the photographs of Ansel Adams, the journals of John Muir, or the adventures of Sir Edmund Hillary. Wild species, not an illuminated Nature, are the components of wilderness. Animals and plants are correlates of our inmost selves in a literal as well as metaphoric sense—literal in the identity of their DNA and our capacity to analogize them as a society.[5]

As for discovering our Pleistocene selves, how are we to translate the question of the hunt into the present? Wilderness sanctuaries presuppose our acceptance of the corporate takeover of everything else. Privatizing is celebrated as part of the ideal of the politics of the state, masked as individualism and freedom. The corporate enterprise in the use of the earth is not interested in either human or natural well-being. Its claims of altruism are made by hired publicists and its sole purpose is to convert the "resources" of the earth into money for its investors. The "trickle-down" benefit for the mass of humanity and for the order of nature is one of the great lies of our time. The last glimmer of a land ethic that adhered to the family farm or the community is vanishing in industrial land use and the corporate enterprise.

✦

WILDERNESS REMAINS FOR ME a problematic theme that is intimately associated in the modern mind with landscape. In this sense wilderness becomes a series of scenes before which spectators pass as they would the galleries of a museum or a kind of scenery for souvenir photographs that we describe to ourselves in a language invented by art critics. Typically lovers of wilderness surround themselves with pictures of mountains or forests or swamps that need not be named or even known.

Art historians attribute the origins of landscape (in the Occident) to fifteenth-century perspective painters, but such pictorializing may have started with Neolithic art at the end of the Pleistocene. Archaeologist N. K. Sandars says: "The tiny size of these paintings is something of a shock after the Paleolithic. . . . The immediate impression is of something happening at a great distance, watched from a vantage-point which may be a little above the scene of the action. This weakens the viewer's sense of participating in what is going forward. There is something of a paradox here, for in the graphic art of the Paleolithic, though man was seldom shown, he was the invisible participant in everything portrayed, while now that he has moved into the canvas and become a principal, there is a quite new detachment and objectivity about his portrayal."[6]

If genre and perspective first appeared in art in the Neolithic, it probably expressed a new sense of being outside nature. Something like modern landscape reappears later in Roman mosaics, prior to its rediscovery by Renaissance artists and their alter egos, the mathematicians, giving evidence of renewed "distancing." It was the same Classical rationality that made possible the lines of latitude and longitude and straight roads across North America, routes based on survey rather than old trails.

To David Lowenthal and Marshall McLuhan I owe thanks for diverting me from writing and thinking about wilderness landscapes as a key to our sense of nature—Lowenthal for his erroneous conclusions about the value of nature being a matter of taste, McLuhan for showing me the connection of that error to the rise of mathematical perspective. With their help I was able to uncover the roots of our present attitudes toward nature, which reach back into European and Mediterranean cultures and the ideology of their organized religions.[7] In their devastating analysis of fifteenth-century science and art, McLuhan and Parker reveal that the emergence of a linear/mathematical perspective and the representation of places framed as pictured objects removed the observer rather than connecting him to his surroundings.[8] Places in pictures could be seen "out there" as through a hole in a wall. This retreat from being in nature is the effect of all the landscape arts—travel, gardening, landscape painting, nature writing—in which we step back far enough to appreciate the esthetic wholeness of the landscape.

Wilderness became a subject matter in art, and the criteria of excellence had to do with technique. The same mathematical referents used by Copernicus and his successors gave us a new vision—explained in such manuals as Leon Battista Alberti's *De Pitura* and carried out by painters like Masaccio and Leonardo—converting the old two-dimensional, asynchronic world (in which different events in time can appear in the same picture) into the Renaissance eye-world composed of repeating units of space and time. The painter Caravaggio is said to have isolated the "moment"—the instant exposure in a temporal world of constant flux. By the eighteenth century the natural world seemed to exist only to end in a picture or, in the other arts, as calculated virtuosity or abstract esthetics.

In this century, the geographer David Lowenthal embodies the insular, humanist position in which the "love of nature" is understood as a "congeries of feelings," a cultural ripple in the wayward tides of fashion.[9] Lowenthal champions a certain educated disdain of existential relevance. He maintains that landscape art is the means of perceiving nature according to criteria established by art criticism by which people "enter" nature as they do a picture gallery. But he misunderstands the truly radical aspect of romanticism by misconstruing it as a play of esthetic symbols rather than a philosophical effort to reintegrate and acknowledge subjective experience, thought and feeling, in art. So long as pictures were regarded as signs and representations arising from the world, landscape could still penetrate all areas of experience, but by the end of the nineteenth century the art world had moved on its trajectory of irrelevance to nonobjectivity, leaving wilderness with the obsolescence and superficiality to which Lowenthal had inadvertently condemned it in the history of taste.

The real landscape has been objectified and distanced through photography, just as the term "landscape" has been misrepresented in literature as a synonym for place, terrain, ecosystem, or environment. Landscape photographs, being surrealistic, empty the subject of intimate context. In time they add layers of temporal distance, leaving a cold crust of esthetics like growing crystals, making the subject increasingly abstract, subjecting real events to the drifting, decadent attention of the gallery coterie and connoisseurs. When nineteenth-century painters discovered photography they were freed, Cezanne said, from literature and subject matter: they could leave the representation of nature to the cameras. But the camera became only another instrument in the pursuit of dissociation, a tool of amputation. In photographs, as the events and people in pictures vanish from living memory, they become only images. Susan Sontag describes this as the emergence of surrealism, disengagement, and estrangement. It is, she says, a separation that enables us to examine dispassionately old photographs of suffering people and to accept the pain and death represented there as though it were less significant than the arrangement of patches of chemicals on paper.[10] With the eye trained by photographs and other pictures to see the land as forms and colors, could judgment of the land itself be far behind?

This schizophrenia—the confusion of reality and fantasy—that the cultured elite seem to value—is a final effect of centuries of splitting art from its origins in religion. Like the painting, the photograph (and eventually visual nature itself) becomes seeing for its own sake—what psychiatrist Bertram Lewin calls "neurotic scopophilia."[11] Photography of nature, which some want to substitute for hunting, must be seen in its true context: pictures of nature representing what is meant by wilderness. Wildness cannot be captured on film; wildness is what I kill and eat because I, too, am wild.

Art set out to simulate and represent nature, probably around fifty thousand years ago, to bring it into the choreography of ceremony and as a context for narrative and mythic referral. It is true that artists' representations often had greater impact on the observer than the original object itself. Attending to content and structure, artists triggered responses similar to that conjured by the natural world. But once art began to be collected, and analyzed in isolation from its original purpose, its moral weight lost its mass. Artists and their patrons created a new language by which intrinsic properties of their work and the virtuosity of the makers were valued. Finally, the abstract characteristics of color, form, movement, symmetry, and line became reality. The next step was the simulation of abstraction itself, in which there was no content to begin with. Painting became a "reflection on the presence of painting itself . . . as if to demonstrate that there are no longer critically reflexive or historically necessary forms with direct access to unconscious truths or a transcendental realm beyond the world—that they are simply *styles* among others."[12] In the end painting represents painting, and any object in nature, any scene or landscape (that is, any *place*), may be taken as an abstraction representing only brushstrokes, signs in which nothing is signified.

One consequence of the abstraction of nature as art is that masses of people, not so interested in arty analysis, now regard the extinction of animals, demise of old-growth forests, pollution of the sea, and the whole range of environmentalist angst as "elitist"—the concerns of the educated affluent who are interested in art in terms of the values of connoisseurship. Nature has been oversold for four centuries as an esthetic as opposed to religious experience. Once associated in the Renaissance with private patrons as part of their affluent playground, art and nature became indulgences of the wealthy.

We hear increasingly about the spiritual uplift derived from a "wilderness experience." The great ecologist Sir Frank Darling describes this spiritual re-creation as "the privilege of the few," adding that "I have an uneasy feeling deep down that we should not burden the wilderness with this egocentric human purpose. The wilderness does not exist for our recreation or delectation. This is something we gain from its great function of being."[13]

✦

TO HIS CREDIT THOREAU did not say, "In wilderness is the preservation of the world." The Great Aphorist did damage enough without confusing wildness and wilderness. Biologist Starker Leopold's research in the mid-1940s on the heritability of subtle vigilance and acute sensibilities confirmed the inherited characteristics of wildness in native turkeys. And, as previously noted in Chapter VI, Helen Spurway, a decade later, provided an effective description of domestication's genetic "goofies" whose ancestors were wild.[14] The loss of wildness in the blunted, monstrous, domesticated surrogates for species, like sanity's mask on the benign visage of a demented friend, is misleading because the original wild plants and animals are gone and cannot be used for comparison.

Wildness occurs in many places. It is composed of the denizens of wilderness—eagles, moose, and their botanical coinhabitants and all of the species whose sexual assortment and genealogy have not been controlled or set adrift by human design or captivity. But it also includes those species who have been cohabitants with domestication—house sparrows, cockroaches, and ourselves. Konrad Lorenz's observations on the bodily and behavioral degradation of domesticated animals, which have been bred to be passive and to have physical conformities that are "babylike" and thus appeal to our protective instincts, show the destruction of wildness.

What then is the wild human? Is it savages? It is us! says Claude Lévi-Strauss—"mind in its untamed state as distinct from mind cultivated or domesticated for the purpose of yielding a return."[15] He refers to the "intransigent refusal on the part of the savage mind to allow anything human (or even living) to remain alien to it."[16] "Wildness does not merely lie behind, it remains the generating matrix," observes philosopher Holmes Rolston.[17] Along with our admirable companions and fellow omnivores, the brown rat, raccoon, and crow, we are not yet deprived of the elegance of native biology by breeding management. The savage mind is ours! We may be deformed by our circumstances, like obese raccoons or crowded, demented rats, but as a species we have in us the call of the wild.

This brings us back to the perennial problem of the other. This is at the core of what it means to be wild. William Arrowsmith observes that in our time "we cannot abide the encounter with the 'other.' . . . We do not teach children Hamlet or Lear because we want to spare them the brush with death. . . . A classicist would call this disease *hybris*. . . . The opposite of *hybris* is sophrosyne. . . . This means 'the skill of mortality.'"[18] It is the obverse side of "giving away"—the way of White Rabbit, in the Lakota myth, reminding the human hunter that he, too, once was a prey and, in terms of cosmic circling back, still is.[19] Wildness, pushed to the perimeters of human settlement during most of the ten millennia since the Pleistocene, has now begun to disappear from the earth, taking the world's otherness of free plants and animals with it. The loss is usually spoken of in terms of ecosystems or the beauty of the world, but for humans, spiritually and psychologically, the true loss is internal. It is our own otherness within.

Wildness and the nature of the self are inextricably joined. Julia Kristeva regards otherness and the self as the deepest problem of civilized life, for it goes directly to self-consciousness. She speaks of two failed solutions: the first is the attempt to transcend the problem by merging with the One—that is, with God—sought in meditation and ascetic solitude by renouncing the physical world: nature is merely the illusion of a mistaken reality. The second is the attempt to see the world as a reflection of the self. Kristeva calls it "the new insanity" of Narcissus. Narcissus, the mythical Greek hunter who is at first beguiled by his own echo and then by his own reflection, ultimately "discovers in sorrow the alienation that is the constituent of his own image." Kristeva speaks of "that lover of himself so strangely close to us in his everyday childishness," but she perhaps does not recognize the failure of the cultural mitigations of neoteny—that aspect of "growing up" which depends on our attention to the natural world and to other species.[20] Likewise, she mentions but cannot explain "the anguish of a drifting mankind, deprived of stable markers," which were in fact the old myths and rites by which the landscape was woven into the spiritual life of primal peoples. The discovery of a psychic self in the ancient world—of an "*inside*, an internal life, to be contrasted with the *outside*"—resulted in isolation of the self, she says, which drove the mechanistic empiricism of the seventeenth century toward "the conquest of the *outside* . . . the outside of nature, to be subjugated by science."[21]

It was also the subjective reintegration of a depauperized outside, shorn of otherness, domesticated, and then internalized as an aspect of the self. We discovered that our own inner otherness—fundamentally perceived as a reflection of the outer forms of life—when bereft of wildness was no longer infinitely mysterious and beautiful and diverse. When Saint Bernard said, "Be ashamed, my soul, for having exchanged divine resemblance against that of beasts," he had the right complaint (falling away) but the wrong subject (wild beasts). He lamented our falling away from abstract heavenly forms rather than from the perfection of the wild world.

Julia Kristeva should not have started with Narcissus, who represents the greedy modern ego. She needed a grounding in an earlier time. She would have done well to read José Ortega y Gasset's *Meditations on Hunting*. In his review of the book James W. Fernandez says that life among hunters "is understood as a mode of coexistence, a dialogue between the subject and its circumstances. Authenticity in life is obtained first by possessing all that is other and second, by turning within, in reflection upon the other, so as not to be mastered by it. . . . The hunter both possesses, or is possessed by, the other through the necessity of imitating the animal in hunting it. But he is carried beyond that possession to the inevitable reflections—meditations—that accompany the death of that other."[22]

The result of such meditation is that the other cannot become part of the self but must remain the alien aspect, a part of our individual being that is not entirely assimilable. It is encountered in other forms of life, the formidable secret of our multiple self, spreading our identity across sixty million years and thousands of species. Sir Thomas Browne puts it this way: "We carry with us the wonders we seek without us: there is all Africa and her prodigies within us."

Domestication is a kind of alchemy whose animals reshape the character of people who have tamed them. Remembering that the opposite of wild is not civilized but domesticated, the best in ourselves is our wildness, nourished by the wild world. To be in a community with crops is to feel like a crop, to have the edges all dulled, our diversity muted. As Konrad Lorenz observed of sheep and domestic rabbits, they are not only dull but mean.

We have been corrupted not only by domestication but by the conventions of nature esthetics. The corporate world has drawn our attention away from wildness with parcels of wilderness that restrict the random play of genes, establish a dichotomy of places, and banish wild forms to enclaves where they may be encountered by audiences while the business of domesticating and denuding the planet proceeds. The savage DNA is being isolated and protected as esthetic relicts, like the vestiges of tribal peoples. The ecological relationships and religious insights of wild cultures, whose social organization represents exotic or vestigial stages in "our history" or "our evolution," are translated into museum dioramas. My wildness, according to this agenda, can be experienced only on reservations called wilderness, but cannot be lived daily in ordinary life.

Wildness should be experienced in the growing of a self that incorporates a person's identity in specific places. To the indigenous people of the Australian outback the terrain is not a great three-dimensional space, not a landscape, but a pattern of connections lived out by walking between places and performing rites that link the individual in critical life stages to sacred places—places that become part of an old story told, sung, and walked through over generations.[23] To be so deeply engaged in place and myth is for most of us today a great hunger.

✦

In our DNA is the wildness that proscribes the limits of a workable physiology as well as a competent culture. Having been generated for over two million years, our wildness requires the taking of life and the eating of flesh of animals as well as fruits and seeds. Life feeds by death-dealing (and death-receiving). The way "out" of the dilemma is into it, a way pioneered for us in the play of sacred trophism, the gamble of sacramental gastronomy, central myths of gifts and chance, the religious context of eating in which the rules are knowing the wild forms who are the game—and being part of the game. You cannot sit out the game except at the cost of health.

Fundamental to wildness is the uniqueness of place, the specific biological niche to which wild species are adapted. Everywhere that "world" religions—Judaism, Christianity, Buddhism, and Islam—have gone, earth shrines, sacred forests, springs, and other places with their wild inhabitants have vanished, replaced often with temples or churches. These "new" architectural structures, often built over an earth shrine and incorporating some of the "old" building materials within them, are often recognizably specific to a certain religion, demonstrating the portability and cosmopolitan nature of that particular religion. We need to ask ourselves if there is a world religion that has a universal philosophy in the sense of inhabiting Planet Earth that, at the same time, demonstrates a consistency with a bioregion and a sense of unique place.

Science and religion have marched together. Since the seventeenth century, physical science has been the secular form of monotheistic abstraction. The writer John Fowles says: "The period had no sympathy with unregulated or primordial nature. It was . . . an ugly and all-invasive reminder of the Fall, of man's eternal exile from the Garden of Eden. . . . Even its natural sciences . . . remained essentially hostile to wild nature, seeing it only as something to be tamed."[24]

Yet nature is so complex that when a correlative of wildness was seen in fluid dynamics, the dismayed physicists cried "Chaos!" The fashionable topic of Chaos represents modern science's consternation with its attempt to "tame" the world: its one-factor approach cannot predict in a world of irreducible variables. Where today's molecules of vapor may be tomorrow at noon no one can say, but not because of the disorder associated with the word "chaos." The attempt to analyze by reduction merely opens an endless series of scales, or "fractal" horizons. It turns out that the closer you look at the edges of things the more they mimic the incredible diversity apparent without magnification. The error is comparable to thinking that a circle is really made up of a very large number of tiny straight lines that would be obvious if only we could magnify it enough. Naturalists knew all along that the world was not chaotic. As genetic mapping inches forward, sometime in the next century the resonance of the two ecologies—the biome and the genome—will be perceived as the key isthmus in the pursuit of human health. The physical sciences will have finally discovered what people already knew, probably for fifty thousand years, that nature cannot be simplified by looking closer and that isolated components have very little value in understanding or prediction. By showing that complexity has limitless dimensions, "the new mathematics of fractal geometry has brought hard science in tune with the peculiarly modern feeling for untamed, uncivilized, undomesticated nature," says James Gleick.[25]

✦

PHYSIOLOGIST RENE DUBOS observes that humans can adapt (via culture) to "starless skies, treeless avenues, shapeless buildings, tasteless bread, joyless celebrations, spiritless pleasures—to a life without reverence for the past, love for the present, or poetical anticipations of the future." But, he says, "it is questionable that man can retain his physical and mental health if he loses contact with the natural forces that have shaped his biological and mental nature."[26] Unless these "forces" are things like "love for the present," what are they? Something "natural" looms behind all this, yet is mediated by culture.

Dubos observes that the human genetic makeup was stabilized about fifty thousand years ago. He quotes Lewis Mumford: "If man had originally inhabited a world as blankly uniform as a 'high rise' housing development, as featureless as a parking lot, as destitute of life as an automated factory, it is doubtful that he would have had a sufficiently varied experience to retain images, mold language or acquire ideas."[27] Our accommodation to ecological dissonance hides our vulnerability to the raw deprivations of cities that we boastfully perceive as elevating us above our progenitors. We have become accustomed to identifying a wide range of physical and social disorders—everything from war to ethnic intolerance, stress and trauma disorders, epidemic disease, and the vague dissatisfactions that lead to addictions and suicide—as weaknesses in the social, political, or technological order, rather than as evidence of a deep, ecological dissociation from our genetic core.

✦

ONE CAN SIMULATE the external features of a primitive life—for example, the limitation of possessions and the nonownership of the land—but "how can the now, by nature explosive and orgiastic, be inserted in historical time?" asks Octavio Paz. *Something* precedes the outward form and its connection to an underlying structure. I have suggested that this something is perception, the way in which sensuous apprehension is linked to the conceptual world, the psychological process by which instinct and ideas interact. The relationship between speech and language is natural and prior to culture, a connection betwixt the palpable world and the conceptual and iconic expressions of it, an event/structure connecting cognition and the outer world, linking gene and environment.

Perception—a precognitive act, mostly unconscious—directs attention, favors preferences, governs sensory emphasis, and gives infrastructure. Perception turns on the central nervous system, which is predisposed to attention and meaning. An example from primitive foragers would be the idea that nonhuman life possesses wisdom, has specific, exemplary powers, is spiritual, and is a mode of apprehension shaped early in life, running deep in our conscious as well as unconscious life. The cycle in the childhood experience of wild things leads to analogical thought and habits of watching nature—a perceptual path taken early that in turn mentors personal growth in a subtle but critical manner. Reconstituting such a cycle is hard. We could get more than glimpses of what is possible if we were truer to our wildness and the intimations from archetypes arising in our dreams or given in visionary moments.

The depth and power of perceptual habits that shape people's lives are described by anthropologist Walter Ong, who distinguishes between an "acoustical event world" and the modern "hypervisual culture." He describes the former as giving primordial design to experience, in which the sound world is more fundamental than the mind's eye.[28] The phonetic alphabet, pictorial space, and euclidean geometry are not just ideas and formulas; they are representations supporting a linear view of the world that in turn shapes our experience of the nonlinear natural world and its creatures. Information based on the reflection of light from surfaces—instead of on messages emanating from the inner life of organisms as is implicit with sound—alters our sense of a living world into a surface world with life sucked out of it.

This view of perception does not mean that we shape our own worlds irrespective of reality. Perception is not another word for taste—or for illusion. In this way, says historian Morris Berman, it transcends " the glaring blind spot of Buddhist philosophy."[29] Perception's truest expression is its contiguity with nature, by which it influences the quality of life, our awareness of ecological integrity, and the connectedness of all things. It is the first step of focused attention and directed awareness. Perceptual habit is "style" in the sense that Margaret Mead defines a group's pattern of bodily movement and sensibility, the predisposition emerging from genetic past and early grounding, affecting every aspect of one's expressive life. In our wild aspect such unconscious expressions are elaborated in dance and narration, surrounded by innumerable and wonderfully varied moral and esthetic presences, nourishing like food that provides both energy and sacrament. Perception provides us with an intuition toward diversity whose forces are sentient, purposeful, and synergistic—a way of expecting, encountering, and experiencing a vast congregation of those unlike us, yet whose shadow reality is part of our deepest self, wild.

Notes

1. The use of "domestic" to allude to warm oatmeal and a nice horse peeking in the door like a Vermeer painting, and all such warm puppy household coziness, has laid a glaze of sugar over the many uses of the term, reducing its usefulness.

2. Notably Niko Tinbergen, Konrad Lorenz, Desmond Morris, Lionel Tiger, Robin Fox, S. Boyd Eaton, Marjorie Shostak, Melvin Konner, Robert Ardrey, and E. O. Wilson.

3. Of course one could add notions like individual freedom or democracy or systems of legal justice. But these are not "ideas" from the last two millennia; they were present long ago in the egalitarian societies of hunter/gatherers.

4. Hugh Iltis, "Flowers and Human Ecology," in Cyril Selmes, ed., *New Movement in the Study and Teaching of Biology* (London: Maurice Temple Smith, 1974).

5. At first I had written "in a symbolic sense." James Fernandez, in his review of Ortega y Gasset's *Meditations on Hunting* (*American Anthropologist* 76(4) [1974]: 868), reminds us what happens when discursive thought attempts to reduce existential complexity, such as hunting, to an essence. A good example, I think, might be the alphabet, the letters of which, as they change over time, move away from any recognizable analogy to things. Ideas tend to be synopticons of a literary tradition, whereas elsewhere they are embedded in things, brought into consciousness by images. See also my book *The Others: How Animals Made Us Human* (Washington, D.C.: Island Press, 1995).

6. N. K. Sandars, *Prehistoric Art of Europe,* 2nd ed. (London: Pelican, 1985), p. 96.

7. Paul H. Shepard, *Man in the Landscape: An Historic View of the Esthetics of Nature* (College Station: Texas A&M University Press, 1991).

8. Marshall McLuhan and Harley Parker, *Through the Vanishing Point: Space in Poetry and Painting* (New York: Harper & Row, 1968), pp. 1–31.

9. David Lowenthal, "Is Wilderness 'Paradise Enow'?" *Columbia University Forum* 7(2) (Spring 1968): 34–40.

10. Susan Sontag, *On Photography* (New York: Dell, 1973).

11. Bertram Lewin, *The Image and the Past* (New York: International Universities Press, 1968), p. 16.

12. Hal Foster, "Signs Taken for Wonders," *Art in America,* June 1986, pp. 80–139.

13. Frank Darling, *Wilderness and Plenty* (London: BBC, 1970).

14. Helen Spurway, "The Causes of Domestication," *Journal of Genetics* 5(3) (1955): 336.

15. Claude Lévi-Strauss, *The Savage Mind,* p. 219.

16. Ibid., p. 245.

17. Holmes Rolston III, "Values Gone Wild," *Inquiry* 26(2) (June 1983): 181–207.

18. William Ayres Arrowsmith, "Hybris and Sophrosyne," *Dartmouth Alumni Magazine,* July 1970, pp. 14–18.

19. Hyemeyohsts Storm, *Seven Arrows,* (New York: Harper & Row, 1972), pp. 340–367.

20. Julia Kristeva, *Tales of Love* (New York: Columbia University Press, 1987), pp. 103–121.

21. Ibid., pp. 376–378.

22. Fernandez, "Review of *Meditations on Hunting,*" pp. 868–869.

23. See Fred Myer, *Pintupi Country, Pintupi Self: Sentiment, Place, and Politics among Desert Aborigines* (Washington, D. C.: Smithsonian, 1986), who is following a path laid out by A. Irving Hallowell in, for example, "Self, Society, and Culture in Phylogenetic Perspective," in Sol Tax, ed., *The Evolution of Man* (Chicago: University of Chicago Press, 1960), pp. 309–371.

24. John Fowles, *A Maggot* (Boston: Little, Brown, 1985), p. 11.

25. James Gleick, *Chaos: Making a New Science* (New York: Viking, 1987), p. 117.

26. Rene Dubos, "Environmental Determinants of Human Life," in David C. Glass, ed., *Environmental Influences* (New York: Rockefeller University Press, 1968), pp. 149–150.

27. Ibid., p. 154 quoted from Lewis Mumford, *The Myth of the Machine* (New York: Harcourt Brace, 1966). Mumford probably got this from Loren Eiseley's "Man of the Future" in *The Immense Journey* (New York: Random House, 1959).

28. Walter J. Ong, "World as View and World as Event," *American Anthropologist* 71 (1969): 634–647.

29. Morris Berman, "The Roots of Reality: Maturana and Varela's *The Tree of Knowledge,*" *Journal of Humanistic Psychology* 23(2) (Spring 1989): 281.

IX

The New Mosaic: A Primal Closure

THREE GREAT COMPOSITE SYSTEMS sustain our lives: the genetic, the ecological, and the cultural. The total heritable material in the individual is the genome, a mosaic of harmonious but distinct entities found in the nucleus of every cell, the result of environmental sifting and selection over a long inheritance. The DNA is organized as linked sequences of genes that constitute chromosomes in twenty-three pairs. Traits are mixed and recombined by sexual reproduction when the germ cells, egg and sperm, coalesce. Because genetic material comes in sets, aberrant recessive genes, although inherited, may remain hidden. Together the accumulation of multiple factors, which recombine randomly, and the mutation of genes, which changes their heritable material, produce great diversity in populations over hundreds of thousands of years.[1]

The structure of the natural community, the ecosystem, is likewise an interdependent whole composed of distinct populations of species in their niches. The biotic community is a composite of linked and yet separable parts, the whole being neither the sum of those parts nor independent of any of them. The ecosystem is the basis of a community physiology, a flow of energy and materials analogous to that passing through a single body. Substitutions can occur: species may be totally removed from a community and a new one may enter—for example, the prairie continued without the buffalo on the one hand and, on the other, survived after the starling arrived from Europe. Ecosystems are not rows of dominoes nor two-dimensional spiderwebs that collapse at a touch. They are living towers of flow that constantly adjust to small tragedies or massive continental shifts.[2]

Not only the genome and the ecosystem but human culture, genetically framed and socially created, is also an integrated and lively conglomerate. Specific art, tools, and beliefs are sometimes gained or lost, moving from culture to culture, carried by people or shared with neighbors. Trailing bits of the context they arrive rough-edged and isolated, but are eventually assimilated as part of the whole.

Genetic systems, ecosystems, and cultures are mosaics that share a common mobility. Genes pass from parent to offspring. Life forms move within and between natural communities by their own power or are carried by other organisms, wind, and water. Cultural elements are borrowed or transported by the migrations of peoples.

Each trait is portable yet embedded, constituting with other bits a whole, a complex, that can be disarticulated and reassembled. The main difference among the three systems is that culture and ecosystems can be changed rapidly, but there is not much we can do about altering our genes. The genes have been selected because they work. They prescribe for society though they do not specify—for example, the genes call for speech but society provides the language. The genome therefore "expects" a certain "fitness" in society and environment. By studying the answer of primal cultures to the demands of DNA, we get the best information about how to construct a human-friendly society and environment. We create ourselves and our world, but our genes dictate the range of feasibility. They specify constraints on our perception of nature and other humans and carry the wisdom of millions of years of selection.

✦

HOW ACCESSIBLE is the Pleistocene? We cannot join the ancestral dead in their finely tuned ecology. But because we have never left our genome and its authority, the strategic nature of the past is born with us. We have only to recognize that our genome and we ourselves are Pleistocene. The two million years (the time comprising our species, *Homo sapiens,* and our immediate ancestor, *Homo erectus*) are crucial for the definition of ourselves: the human genome is the blueprint that frames our choices of ways of life, of healthy or sick cultures. Guided by our genome, we can respond appropriately and consider thoughtfully the options for living in this modern world that will take us home to our primal intuitions and needs.

While hunter/gatherers have been studied mainly for their differences from modern humans, they share many similarities and it is these we may take as guideposts. Many of the common traits are social, framed within what might be called "the fire circle." Gathering around a fire is one of the deepest images of our collective memory. From as long ago as 700,000 years we have met around fires. The fire circle embraces us, socially and culturally, even today across aeons of time. Symbolically it is something we understand, the cement of an extended family and community structures. Much about being human is thereby signified: the social unity of the small group, the sharing of food and understanding, the anticipated collaboration in foraging, repose, reflection, and solidarity that could only come where the lifelike flame reached out as though from a common heart. We have progressed far in ecological thinking. We now need to develop an "ecological civicism," as Claude Lévi-Strauss has suggested—one that restores the organic bonds of community. The essence of the fire circle should be maintained in our modern world as the central metaphor for such human gathering and sharing.

The greatest innovation of humankind was not the employment of oil and coal but the constituting of that fire circle. Socially the fire circle is typically composed of about twenty-four persons or a twelve-adult council of the whole. This group size is magic for our species. It is deeply embedded in the human unconscious, the perfect number to deal with a problem, visit and dance, mourn and celebrate, tell a story, plan for tomorrow, hold a council, or eat a bison. Fire was perhaps the first metaphor and therefore the master stimulus to deliberation, the symbol of life itself. It is both literally and symbolically the place where the occasions of the life cycle are met with enfolding care and customs traditional to the group.

In his or her lifetime, a person moves through physical and psychological states that match both bodily transformation and images of the world which will become a spiritual vision that makes one at home in the world. At night around the forty or fifty feet upon which the light of the flame dances, all else is invisible and the group is brought together sharing warmth, light, smell, and sound. The horizon closes to make a small world. But with the opening of the night sky there is a new expansion. The middle world of the terrain and the underworld of deep water vanish and in their place is the overworld and the combined social being of people and their fire. The fire circle engages those no longer present even in this modern world where we so often have forgotten our ancestors. Attention to the "presence" of ancestors and to kinship is fundamental to our personal identity and our understanding of the ecological ties of all creatures, past and present, on this Planet Earth.

The fire circle is where the day's foraging comes to a close and begins. The day's events are reviewed and sometimes pondered and the next day is considered. The small group works better for its members as a functional institution.[3] The fire circle emphasizes the importance of the small-scale in human affairs. All forms of social connection are signified in the reach of the voice, the vibration of dancing feet, and the light that penetrates into the dark surround. Intertribal tension-reduction formalities, such as song duels, peace-pipe ceremonies, and competitive sports tend to be aligned with the fire circle, where disagreement and fighting, as among animals, become symbolic.

In the face-to-face decision making around the fire circle, adult power is sharply limited. Even freely elected representatives in a democratic society are poor substitutes for this kind of decision making open to all. Hunter/gatherer groups are usually described as making decisions as a council of the whole in which all have freedom to speak. Among primal peoples there is no representational government. The participant voice, not higher politics, is typical of small groups and virtually impossible in large-scale societies.

Prestige is based on ability demonstrated, through persuasion and example. Dynamic, emergent, and dispersed leadership is not monolithic but divided and is not predicated on conflict, competition, or submission. Again and again, leadership in small groups has proved to be keyed to integrity rather than imposed or inherited authority. Only in agricultural "big chief," "headman," or larger societies is power inherited, delegated, or obtained by intrigue. Centralized power is the great curse of history.

✦

In earlier chapters I discussed the importance of ontogeny—a resonance between bonding and separation that produces confident selfhood supported by periodic acts of public recognition--and the significance of neoteny, the retention of immature traits, in this process. At this point it is important to recapitulate, in terms of cultural consequences, the necessity of properly attending to these processes in our children and youth in the manner of our Pleistocene ancestors. I have shown how neoteny is the extraordinary human biological adaptation, a purposeful retardation, to which the life cycle is attuned and to which the culture must append its calendars. Neoteny creates a vision of the world based on a perceptual developmental agenda in which the individual is enfolded successively by mother, by nature, and by the universe.

The dynamics of bonding and separation are played out in these successive stages between the infant and parent, the child and nature, the young adult and the cosmos. Life is appropriately lived through a series of "mothers." It is focused in ever widening spheres of experience and identification in which the self emerges as a constituent of the social, ecological, and spiritual environments, each organically modeled.

Our ontogeny prepares this expansion, demands it, but culture must answer. In tribal life there is usually formal group acknowledgment and celebration of the individual's life stages and passages. Many of these stages are marked by visible "biological" signs: ceasing to nurse, walking, talking, tooth replacement, puberty, hair growth, various skills, parenthood, hair color changes or hair loss, wrinkles and other signs of maturity: intellectual, psychic, and recollective. The key nurturant occasions are triggers in this essential epigenesis. The mitigation of our valuable retardation is episodic and social, requiring a matching of the calendars of development through infancy, childhood, adolescence, and adulthood with keen awareness and appropriate responses by caregivers, mentors, and society—and mediated by closeness to nature, our primary mother. Our genome "expects" society to act in this tutoring with a testing and ceremonial response to the personal agenda; otherwise we slide into adult infantility and its neurotic symptoms. And when individuals are prepared to move into adulthood keenly aware of their fundamental continuity with the natural world, the ecosystem too benefits from good tutoring by the whole social community. This astonishing arrangement is foreshadowed in a design provided in the nucleus of every human cell. It is an expectation of the genome, fostered by society, enacted in ecosystems.

Cultural practices—the stories, rites, skills, art, and social markers of status—take the novitiate through early life, step by step, stage by stage, each stage freeing up the next: a fledging and moulting principle. Long life and long youth are a great gift, but they subvert personal identity unless marked by progressive, epigenetic feedback that couples gene and environment through social acts. Genes and experience are interrelated and finely tuned by the complex biological specialization of our slow or "retarded" development to which human culture is mediator. We do not become more mature by thinking ourselves into it. Only *culture* can mitigate neoteny. That is to say, a group of our friends or relatives guides us and responds to us. Otherwise we remain psychically inchoate.

Being individually slow to reach maturity, we are among the most neotenic or babylike of species. Biologically we invest our reproductive energy in few offspring and their slow individual development and education. Slowly culture fills in the gaps in our development and mitigates our incompleteness according to an inherent timetable. Incomplete ontogeny simply grinds into the dead ends of infantility and pathological limbo.[4]

Two of the transformative stages of human ontogeny have been studied in detail among living hunter/gatherers: infant/caregiver relationships and adolescent initiation. The archaeological record leaves little doubt that these are ancient patterns which may be incompletely addressed in ourselves. Foremost is the bonding and separation dynamic of the first two years. Childhood among hunter/gatherers better fits the human genome in terms of the experience and satisfaction of both parents and children than it does in our own time.[5] The interaction of infant and caregivers emerges as a compelling need—perhaps the most powerful shaping force in individual experience. Oddly enough, bonding's "purpose" is separation, successive steps of coming together and departing, in which the individual emerges in new relationships to humans and nonhumans. Details of the socially embedded rhythms of parenthood vary from culture to culture, but they can hardly improve on the basic style or primary forms found in hunter/gatherer groups. Intense early attachment leads not to prolonged dependency but to a better-functioning nervous system and greater success in the separation process.[6]

The "social skills" of the newborn and the reciprocity of the mother, father, aunt, uncle, grandparents, and siblings create not only the primary social ties but also the paradigm for existential attitudes and the nature of Others. The lifelong perception of the world as a "counterplayer"—as either caring, nourishing, instructing, and protecting or else as vindictive, mechanical, dangerous, or distant—begins here. The process arises in our earliest experience and is coupled to patterns of response. Hara Marano says: "Newborns come highly equipped for their first intense meetings with their parents, and in particular their mothers. . . . Biologically speaking, today's mothers and babies are two to three million years old. . . . When we put the body of a mother close to her baby, something is turned on that is part of her genetic makeup."[7] The crucial factor in family organization is the way in which children are reared. Extended lactation and nursing are both biologically protective and psychologically beneficial. Not only does this have emotional and physiological benefits for both mother and infant, it also helps regulate the population through hormonal control of ovulation, often rendering the mother infertile for its duration.

Children in primal societies have a richly textured play space and earth-crawling freedom in infancy. This zone extends from a circle around the infant's mother or other caregivers to the small world of the juvenile with its own terrain, plants, animals, and artifacts, ideally including trees and water. The regular opportunity of the child while yet in the care of others to move about on the ground amid plants, to taste the earth and engage its bacteria, creates a sensory and chemical atunement to place: a kind of imprinting.

Children in primal societies have access to the scenes of life—such as butchering, copulation, birth, and death—especially within the family and in nature. They live in a rich, nonhuman plant and animal environment at the time of language acquisition and are given the opportunity to name animals with a coplayer. Taxonomy is fundamental to cognition as well as to grounding in a real world. From birth the lives of children are keyed to the daily, monthly, and seasonal round. These cycles are the true pulse to which their blood resonates, as distinct from the clock, electronic calendar, and historical regulators of our own lives. In this way the lifetimes of children are seen as part of other periodic natural events.

Games, based on animal-mimic play and other introjective predications of animals on the "inchoate" individual, are natural to children. Pretend play is the internalizing of the living world to create an enacted and then a perceived orderliness in the self that includes the verbing of animal names and the use of animals as models of special skills. In human small-group society this nonpeer play is unlike most of our school and recreation groupings in which children are classed by age. The concept of the game is a homology of the hunter and his quarry—of which the hunter is the most profound student and venerator, and the prey, the opponent, is equally fervent. To love and not to hate the opponent must be understood as a spiritualized expression of life.

Toys in modern society may be a burden to children in ways we do not yet understand. Toys are precursors to material possessions, which are few among primal peoples. They objectify the world as passive and subordinate to ourselves and, despite childhood pretending, are nonliving. Toys may be symptomatic of social deprivation, solitude, and isolation.

The importance of cultural support does not end with childhood but continues as the individual begins sexual development during puberty. During this transition to adulthood, youths need continued understanding and thoughtful guidance by the adult community. Adolescent initiation—and its importance to the development of the individual and his or her integration both into the group and into a cosmos—is of great significance in tribal cultures. Among many groups there is a formal initiation into a religious body during which the novitiate temporarily loses identity and is reborn. Much of modern angst has its roots in the episodes of failure to mitigate regressive psychological traits in personal development between the twelfth and sixteenth years . . . and the endless social aggression that follows. The lack of appropriate initiation ceremonies for adolescents today is a glaring tear in the fabric of society, patched up by sports, teams, and clubs and exacerbated by gangs, where adolescents create their own identity without the watchful guidance of elders.

Aging is inevitable, but growth beyond childhood and adolescence, through successive levels of leadership and social responsibility, depends on the embrace by the human community. So-called postreproductive advisory functions—that is, grandparental roles—are fundamental to human society, particularly in advice to young parents, but also in recollecting social experience, clan and family history, and knowledge in matters of plants, animals, and places. Elders are a special generational repository for the mnemonics of recollection. Likewise, the nonreproductive members of society (the unmarried, widows and widowers, and homosexuals) play unique roles that may not be immediately apparent but are especially valuable in their watchfulness of the puzzling young who otherwise grow narrow in their intolerance of others. They are the keepers of the old stories and old ways as well as models of alternative modes of living.

Clan and other memberships progressively support the development of identity with age. The extended family is of fundamental importance to human sanity. Aunts and uncles, grandparents, cousins, and in-laws are necessary but inadequately appreciated in modern societies like ours. Aunts and uncles are essential members of the primal household because they are halfway between parents and others and therefore sometimes in positions of more authority and less conflict than parents.

A tribe, constituted by a genetic/marriage/linguistic grouping, is probably the best size—best for a sense of belonging, for the freedom for fission by individuals or families in smaller bands, and for the maintenance of genetic diversity. The magic numbers for size of tribes, clans, and extended family, when we learn to understand them, will undoubtedly reflect optimum groupings for human becoming according to our nature as a biological species.

If we reach decisions more effectively in small groups, and if cultures are more flexible and innovative in groups of a few thousand, it is probably because we evolved in those terms. In such groups there is limited exposure to strangers—who are always considered "other" but are met hospitably and with ceremony at boundaries, or invited as transients or guests, but rarely become accretions to the community.[8] Like other terrestrial primates, humans sometimes enjoy larger congregations. Such gatherings seem to be part of a renewal process that affirms the value of the exogamic (out-marriage) mixing of linguistic and racial structure.

✦

PLEISTOCENE HERITAGE can inform us about the sources of our ecological problems—especially regarding health and well-being. Health disorders today are increasingly traced to polluting poisons, domesticated (that is, chemically altered or chemically treated) plants and animals, and the chemical changes in our bodies resulting from increasing cholesterols, fats, and carbohydrates in our diets. Food for tribal peoples includes a mix of wild-gathered seeds, nuts, plants, and small animals with fresh or dried garden produce, as much eaten raw as possible. An ecotypic economy of this kind is keyed to place—using local varieties of plants or breeds of animals, and gathering and hunting the familiar life forms.

A foraging society is not one in which a particular group or gender is more kind, moral, ethical, or informed than another. Among hunter/gatherers there is no religious minority in the form of a Great Mother or goddess cult. The modern attempt to associate feminism, vegetarianism, and animal liberation in any historical or anthropological framework is unfounded.[9] Primal society is not grounds for a new "me first" between women and men or between vegetarians and meat eaters. The human digestive system and physiology cannot be fooled by squeezing a diet from a moral. We are omnivores: our intestines and teeth attest to this fact.

Before we ran we climbed and before that we crawled and swam. Running was the act in the final making of our species' body, and now we must run to be well, climb for excitement, and swim for the satisfaction of the torso. Regular exercise, especially jogging, aerobics, swimming, and stretching, correlates with certain routines of life in hunting societies whose benefits are not only physical but mental.[10] The immense literature on the benefits of exercise needs no review except to point out that such activities are healthful for all peoples because our bodies were designed through millions of years of vigorous exercise.

Patterns of energy flow were the great metaphysical discovery of the prehistoric world because they were analogies for social life and for the structure of the whole cosmos. The Neanderthals observed in bears—who went into their deathlike hibernation and reemerged in the spring with new vigor and often new little cubs at their sides—the model of the cycle of life, given and recovered. Energy flow is fundamentally the same to our life processes as it was for our distant cousins. Aldo Leopold's story of the passage of an atom from a dead buffalo through decay, the chain of photosynthesis, predation, and back into organisms and mineralization retells the old tale of death and the return to life in ecological terms.

Animals on the medicine wheel of the Plains Indians were said to be those that know how to give away. "Each dot I have made with my finger in the dirt is an animal," said White Rabbit in a Lakota myth. "There is no one of any of the animals in this world that can do without the next. Each whole tribe of animals is a Medicine Wheel, in that it is the One Mind. Each dot on the Great Wheel is a tribe of animals. And parts of these tribes must Give-Away in order that they all might grow. The animal tribes all know of this. It is only the tribes of People who are the ones who must learn it."[11]

Gary Snyder reminds us of this ancient story: "Everything that lives eats food, and is food in turn. . . . To grossly use more than you need, to destroy, is biologically unsound. Much of the production and consumption of modern societies is not necessary or conducive to spiritual and cultural growth, let alone survival; and is behind much greed and envy."[12]

As for the killing, only ceremony and a profound belief in the sentience and spirituality of all things allow one to deal with death. Ceremony affirms the ultimate transformative processes of birth and death, preserving openness and difference. The hunter's idea is that of "being played on like a pipe."[13] After the killing of an animal there is a stillness when thoughts of life's brevity and preciousness are present.[14]

Recent attempts to create male fellowship bonds are evidence of a profound emptiness in the lives of men. But fellowship and maturity have nothing to do with a warrior brotherhood, a kind of arrested juvenility. Hunter, warrior, soldier are not a continuum except in a historically destructive sense. Of all the scores of tools in the great Paleolithic atelier, there is not a single weapon designed for war. There is no antecedent for a state of war in Pleistocene primal groups.

As we saw in Chapter VII, the warrior is the culturally degenerate hunter cast in the pastoral authoritarian system. The warrior came into existence with the herding of domestic ungulates. In his earliest proto-Mesopotamian existence the warrior was armed from the hunt and reoriented to kill people and steal cattle. His highest ideals—loyalty and obedience to a superior—were a kind of defect of true fidelity, which Erik Erikson defines as "that virtue and quality of adolescent ego strength which belongs to man's evolutionary heritage, but which—like all the basic virtues—can arise only in the interplay of a life stage with the individuals and the social forces of a true community."[15] James Mischke, a teacher and counselor at Déné College, describes the great destruction of the original Navajo foraging culture and the rise of internecine tribal wars following the acquisition of goats, sheep, and cattle. The rise at that time of the hero/warrior, he says, was far more disastrous for Navajo society than the advent of colonial militarism two centuries later.[16]

Today we cannot become hunter/gatherers as a whole society, but we may recover some social principles, metaphysical insights, and spiritual qualities from their way of life by reconstructing it in our own milieux. The hunt brings into play intense emotions and a sense of the mysteries of our existence, a cathartic and mediating transformation. The value of the hunt is not found in repeated forays into the outback but in a leap forward into the heart-structure of the world, the "game" played once by rules that now illuminate our real selves.

The modern heaping of abuse on hunters who "kill things" has escalated in the last years of the twentieth century. This rising hysteria about killing reminds us that the problem with death was never so intense among primal peoples—who participated in the great round—as it is among those societies who dread it as a final calamity and strive to deny it and for whom it becomes a neurotic obsession. Psychological research indicates that moral codes are more rigid among people who dread death and whose inflexibility is projected into all kinds of social conservatism.[17] When morality is premised on the escape from death, it is aimed at all those "causing" or participating in it. Most death in nature is invisible and, moreover, is accompanied by the fantasy that animals who are not killed (by people) go on living. The killing of one animal by another, so seldom seen, can be ignored or turned back into the unconscious. This repressed notion about nonhumans is released as fury against human hunters. The gore of carnivory, the predatory bite, the lethal stealth of the parasite, the decimation of wild "babies" by hyenas or skuas, the death throes—all bloodshed and butchering seem horrible when crudely anthropomorphized.

The decade of the 1980s witnessed a spate of essays on the "morality" of hunting.[18] Focus on the ethics of hunting decontextualizes the subject. Its rhetoric of killing as evil, and compassion as its opposite, is abstract. Animal rights ethicists disembowel the subject the way a small mammal is collected for taxidermy. Having taken away the guts that connect the animal to its surroundings, there remains a shell, deprived not only of its own life but of the putrefaction as well that reintegrates the dead with the living.

✦

CEREMONIES, DANCES, ART, AND STORIES are ways of recalling. Genes are not only "how-to" information but bodily memories of past environments and responses to them. The reconciliation of our recollected selfhood with our genetic basis transcends the dichotomy that defines us today.

Narrative and sequential recall are basic to human thought. Very probably they began with recitation of the hunt with gestures and pantomime as part of a lively retelling. Alexander Marshack speaks of "the demands of fire culture" in which the tale is a "metaphysical gift" making the world "an object of contemplation."[19] The first art in human culture may have been the oral story. All great tales are in some sense recapitulations. Robert Ridington says: "When the man and animal do meet it is a moment of transformation, like the moment of meeting in the vision quest when the child enters the animal's world of experience and is devoured by another realm of consciousness. . . . The vision quest symbolically transforms the child's meat into spirit, and the hunt transforms the animal's spirit into meat."[20]

Among seminomadic hunting/gathering peoples there is little accumulation of personal goods, minimal housekeeping, and few structures. With foraging peoples, tools and other personal objects of apparel or adornment are hand-crafted and mostly used by the maker, or within the family, or are sometimes bartered. Because their manufacture involves stereotyped hand movements, tools are a kind of solidification of gestures and a representation of the personalities of the crafters. The Paleolithic or "Old Stones" age is not central to the origin of craftmanship, since all hunters and gatherers have made more use of organic material than of flint. It just happens, however, that the evidence from stone—in artifacts, pictographs, and petroglyphs—is our best source of information on the prehistory of speech, art, and narration.

What seems to us a chancy way of life must somehow be understood as an ongoing arrangement in which we participate, though not as masters. To extract material rewards from the world through strategy is necessary, but conniving is less important than being right with the deities. Diffused sacredness, a strong sense of transformation, and unhistorical time constitute the Paleolithic genius. As ideals not one of these is a regression into archaic obsolescence but a forward step to modern philosophical thinking.

Art should return to its roots, to cosmology, to rite, and to ceremony. The religious nature of art is its true meaning. Modern art's commitment to "emotion" and "feeling" or to abstract principles of design is, by Pleistocene standards, a sacrilegious act, just as narcotics belong not in a recreational but in a religious setting. In most small-scale societies there is regular dialogue on divinatory and dream experience that gets translated into art.

In her book *Prehistoric Art of Europe*, N. K. Sandars identifies the primordial human experience of the divine: "the sense of diffused sacredness which may erupt into everyday life . . . an order of relationships the categories of which take no account of genetic barriers and which will lead to ideas of metamorphosis inside and outside this life . . . unhistorical time . . . and the character or position of the medicine man or shaman."[21] To this we should add that sense of a giving or gifting world in which the improbability and yet inexplicable provision for life can be appreciated.

✦

ECOLOGICALLY, the Pleistocene asks that we free ourselves from domestication. In a better (but not other) world there would be no monocrop agriculture, hybrid seeds, chemical fertilizers, or industrial pesticides. An escape from domestication would liberate nature into itself and free us from the tyranny of the created blobs and the emotional stuckness of ethical humanism and agrarian brutality.

The physical damage to the earth of our cultivations and grazings is widely known. Fundamental damage by domestic animals to the human psyche is twofold: it demeans and destroys the meaning of wild species by substituting rough, inferior copies for reality, subverting a true biophilia; it injures the perceiver, too, by granting him powers over the animal and a kind of ersatz familial responsibility that become part of the human personality.

A geographic orientation that is continuous with the passages of life and with special places connects the foragers with the mythic origin of people and with important occasions. Landholding as an end in itself is unnecessary when human numbers are small and nature is a shared "storehouse." Primal peoples do not own land and evince little absolute territoriality.[22] The terrain is a commons and movement is on foot.[23] Outsiders are admitted as a courtesy when they observe the appropriate protocol. Plurality is accommodated rather than denied. Thus did the "peace pipe" among American Indians serve its mediating role.

Space in our society has largely replaced Place, and this is a great loss for our children. Our experience as children of particular environments is part of that identity-forming process basic to our sense of self, generated over millions of years of inhabiting a home range with its unique features. Roughly the child's space is measured by the range of the mother's voice. Place is at once an external and internal state in a journey home—a process based not on mathematical coordinates but on specific geology, climate, habitat, and inhabitants. Terrain is analogous to the self.

Ecology is largely conceptual. One sees plants and animals, the terrain, water, and sky, not "an ecology." Animals and plants are the language of nature, together participating in human perception in a great semiosis, a principle of analogy and a gift to human society. Before the signs turned into an alphabet, we read the world as the hunter/gatherers read tracks in a world of metaphors of human society, a special analogy to ecology. This kind of savage "nature study" is the avocation of many foraging peoples but imposes nothing.

Raymond Chipeniuk, a professor of regional planning and resource development at Brock University in Ontario, Canada, has conducted research with schoolchildren showing that "foraging for natural things in childhood develops competence in assessing biodiversity of local habitats.[24] He says that "domain-specific" components of the human mind, "modes of cognition about nature, ways of thinking, [are] governed by rules that are not specific to culture."[25] All children think in fundamentally different ways about natural things than they do of made things—mainly that there are unseen essences that govern natural beings while artifacts are seen as simply collections of characteristics that define things in terms of utility. The taxonomy of natural forms is structured differently from that of made things. "Conservationists are actually doing harm," he says, "by trying to convert the lay mind to the view that human artifactual reality is part of nature. . . . We must make ourselves indigenes, natives returning to a lost landscape still somewhere in our hearts and minds, the Paleolithic ideal, a country of abundant wildlife, rich and fruitful native vegetation, pure water, and fresh air. Otherwise we will remain what we have been for ten thousands of years: a sort of perpetual exotic, never ceasing to invade, disrupt, and degrade the pre-existing natural landscapes." Culture is not inherited via the genes but, according to Chipeniuk, there is "an inborn and cross-cultural preference for naturalness in landscapes."[26]

People respond differently physiologically to built and natural landscapes, favoring especially savannas. This is natural habitat selection by a creature whose most definitive millennia were spent searching grassland and forest edge. This sensibility is important to us today in assessing the health of our environments. Chipeniuk proceeds to itemize the value of foraging expeditions by children, which culminates in their ability to see how the world is going. He points out that gathering is still widely practiced in Europe and America and that most of the great naturalists were also hunters. The ideas of naturalness and foraging as a better learning experience can be seen by anyone who has witnessed the short-lived results of "teaching modules" about the "ecology of the rain forest" in contrast to going out and studying one's home area.

✦

OUR HUMAN ECOLOGY is that of a rare species of mammal in a social, omnivorous niche. Our demography is one of a slow-breeding, large, intelligent primate. To shatter our population structure, to become abundant in the way of rodents, not only destroys our ecological relations with the rest of nature, it sets the stage for our mass insanity.

Discussions about human population usually center on physical resources, but the true problem is much worse. When all land looks like much of the world between the Tropics of Cancer and Capricorn, it will be too late. As ethologists Konrad Lorenz and Paul Leyhausen have observed, "Space . . . is indispensable for the psychological and mental health of humans. . . . Overcrowding is a menace to mankind long before general and insurmountable food shortage sets in. The increase in human numbers is not primarily a food problem, it is a psychological, sociological, mental health problem. . . . We have to realize that human nature sets a far narrower limit to human adaptability to overcrowding than is commonly believed."[27]

The world is full of war, terrorism, social disintegration, poisoned air and land. The soil that has accumulated for centuries is washing into the sea; the earth's forests are being devastated. Virtually all the diseases of the past are with us in more virulent form, and new epidemics of psychic breakdown, dysfunctional families, and organic infection are upon us. The last benefits of the raiding of the earth by the affluent minority still give us an illusion of well-being in the midst of worldwide calamity.

How may we recover the experience of a world in which we are surrounded by a multitude of conscious, powerful beings, incarnate as natural forms? No act of will seems sufficient. We have been ideologically deflowered. The demythologizing set in motion by the ancient Greeks and Hebrews seems overwhelming. Powers outside one's self—perhaps the "collective unconscious" of Carl Jung, preconscious perception of the rightness of primal life—may be at work. What is said and done and shown around infants and children affects their later cognition. Archetypal imagery represents the real, past world. It did not "get in" our brains by chance but as some profound process of human assimilation. The effort to recover an appropriate response to it seems frivolous only because we have grown up narrowly committed to rational experience.

✦

A JOURNEY TO OUR PRIMAL WORLD may bring answers to our ecological dilemmas. Such a journey will lead, not to an impulsive or thoughtless way of life, but to a reciprocity with origins declared by history to be out of reach. When Ortega y Gasset speaks of hunting as "a deep and permanent yearning in the human condition," our "generic way of being," he refers to the whole of the foraging way of life, which we can shape in detail to our own time.

We live with the possibility of a primal closure. All around us aspects of the modern world—diet, exercise, medicine, art, work, family, philosophy, economics, ecology, psychology—have begun a long circle back toward their former coherence. Whether they can arrive before the natural world is damaged beyond repair and madness destroys humanity we cannot tell.

What the West has going for it is the tradition of self-scrutiny, self-criticism, and access historically and scientifically to other cultures. The human psyche makes unremitting demands for physical and spiritual (or symbolic) otherness, and the modern West has the information if not the wisdom for escaping the trap of industrial productivity, corporate blight, and demographic insanity.

We can go back to nature, as I wrote in 1973, because we never left it. To illustrate this I have formulated some seventy-odd themes of cultural recovery selected from the record of primal cultural traits as played out over thousands of years (Table 2). It is time to abandon the fantasy that we are above the past and alienated from the rest of life on earth. We truly are a successful species in our own right that lived in harmony with the earth and its other forms for millions of years—a species that has not changed intrinsically. The genome is our Pleistocene treasure that transcends short-term and short-sighted goals. Possibilities lie within us. Our culture must express what the past calls forth in us but leaves us the freedom to shape.[28]

To reenvision "going back," we look with our mind's eye at time as a

TABLE 2. ASPECTS OF A PLEISTOCENE PARADIGM

Ontogenic	
1.	Formal recognition of stages in the whole life cycle
2.	The progressive dynamics of bonding and separation
3.	Earth-crawling freedom by 18 months
4.	Richly textures play space
5.	No reading prior to "symbolic" age (about 12 years)
6.	All-age access to butchering scenes
7.	All-age access to birth, copulation, death scenes
8.	Few toys
9.	Early access via speech to rich species taxonomy
10.	Formal celebration of life-stage passages such as initiation
11.	Rich animal-mimic play and other introjective processes
12.	Non-peer-group play
13.	Parturition and neonate "soft" environment
14.	Access to named places in connection with mythology
15.	Extended family or dense social structure
16.	Extended lactation
17.	Play as the internal prediction of the living world
18.	Little storage, accumulation, or provision
19.	Diversity of "work"
20.	Handmade tools and other objects
21.	No monoculture
22.	Independent family subsistence plus customary sharing
23.	Ecotypic economy—keyed to place
24.	No landownership in the sense of "fee simple"
25.	Little absolute territoriality
26.	No fossil fuel use
27.	Minimal housekeeping
28.	No domestic plants or animals
Social	
29.	Prestige based on demonstrated integrity
30.	Little or no heritable rank
31.	Size of genetic/marriage/linguistic group or tribe: 500–3000
32.	Clan and other membership giving progressive identity with age
33.	Limited exposure to strangers
34.	Hospitality to outsiders
35.	Functional roles of aunts and uncles
36.	Postreproductive advisory functions such as grandparental roles

(continues)

TABLE 2. ASPECTS OF A PLEISTOCENE PARADIGM (*continued*)

37. Size of fire-circle group: 10 adults (council of the whole)
38. Occasional larger congregations
39. Emphasis on mneumonics with its generational repository
40. Participant politics vs. representational or authoritarian
41. Vernacular gender and age functions
42. Totemic analogical thought of eco-predicated logos
43. Dynamic, emergent, and dispersed leadership
44. Decentralized power
45. Intertribal tension-reduction rites (song duels, peacepipe)
46. Cosmologically rather than sociohierarchically focused ritual

Other

47. Periodic mobility, no sedentism
48. Conceptual notion of spirit in all life, numinous otherness
49. Centrality of narrative, routine recall and story
50. Dietary omnivory
51. Rare-species demography
52. Subordination of art to cosmology
53. Participatory rather than audience-focused music
54. Sensual science ("science of the concrete") vs. intangible science
55. Celebration of social and cosmological function of meat eating
56. Religious regulation of the special effects of plant substances
57. Extensive foot travel
58. Only organic medicine
59. Regular dialogue on dream experience
60. The "game" approach —to love, not hate, the opponent
61. Attention to listening, to the sound environment as voice
62. Running
63. Attention to kinship and the "presence" of ancestors
64. Attunement to the daily cycle and seasonality
65. No radical intervention on fetal genetic malformations
66. Immediate access to the wild, wilderness, solitude
67. Nonlinear time and space—no history, progress, or destiny
68. Sacramental (not sacrificial) trophism
69. Formal recognition of a gifted subsistence
70. Participation in hunting and gathering
71. Freedom—to come and go, to choose skills, to marry or not, etc.

spiral rather than a reversal. We "go back" with each day along an ellipse with the rising and setting of the sun, each turning of the globe. Every new generation "goes back" to forms of earlier generations, from which the individual comes forward in his singular ontogeny. We cannot run the life cycle backwards, but we cannot avoid the inherent and essential demands of an ancient, repetitive pattern as surely as human embryology follows a design derived from an ancestral fish. Most of the "new" events in each individual life are like a different pianist playing a familiar piece.

White European/Americans cannot become Hopis or Kalahari Bushmen or Magdalenian bison hunters, but elements in those cultures can be recovered or re-created because they fit the heritage and predilection of the human genome everywhere, a genome tracing back to a common ancestor that Anglos share with Hopis and Bushmen and all the rest of *Homo sapiens.* The social, ecological, and ideological characteristics natural to our humanity are to be found in the lives of foragers. As I have suggested, they are our human nature because they characterized the human way of life during our evolution.

Must we build a new twenty-first-century society corresponding to a hunting/gathering culture? Of course not; humans do not consciously make cultures. What we can do is single out those many things, large and small, that characterized the social and cultural life of our ancestors—the terms under which our genome itself was shaped—and incorporate them as best we can by creating a modern life around them. We take our cues from primal cultures, the best wisdom of the deep desires of the genome. We humans are instinctive culture makers; given the pieces, the culture will reshape itself.

Notes

1. Through this sorting process, neoteny is made possible. Some traits that are present in immature organisms but adaptive for mature individuals are carried over through adulthood, other adaptive traits are retained; and some nonadaptive characteristics are deleted from populations. Thus in humans the shoulder girdle and arm represent old adaptive traits whereas the arrangement of the leg and pelvis represent relatively new ones in terms of evolution. In this way, over long periods of time, changes in genetic traits occur as organisms with adaptive traits survive and those with maladaptive traits die out.

2. The massive destruction of ecosystems, which may take thousands of years to recover, is not the topic here.

3. Jane Howard, "All Happy Clans Are Alike," *Atlantic*, May 1978, pp. 37–42.

4. Arnold Modell, "The Sense of Identity: The Acceptance of Separateness," in *Object Love and Reality* (New York: International Universities Press, 1968), pp. 43–62.

5.Melvin J. Konner, "Maternal Care, Infant Behavior and Development among the !Kung," in Richard B. Lee and Irven DeVore, eds., *Kalahari Hunters and Gatherers* (Cambridge, Mass: Harvard University Press, 1976), pp. 218–245.

6. Ibid.; see also Patricia Draper, "Social and Economic Constraints on Child Life among the !Kung," in Lee and DeVore, *Kalahari Hunters and Gatherers*, pp. 199–245.

7. Hara Estroff Marano, "Biology Is One Key to the Bonding of Mothers and Babies," *Smithsonian*, February 1981, pp. 60–68.

8. In this regard, it is interesting that alienation and indifference in modern society usually involve socially neutral situations and strangers. This knowledge should lead us, in law and education, to look with understanding at problems associated with integration of races and religions and to examine our own innate fear and hatred of those "others."

9. For example, the stridency of the opposition to hunting suggests that it is motivated by the denial of death. The trophy hunter is widely ridiculed for his wall covered with the mounted heads of his quarry. Yet even this is part of a tradition of venerating the animal by special attention to the head—one of the oldest continuing customs in human life whose antecedents have been carbon-dated at fifty thousand years. Such traditions are, except where corrupted, surely a survival of something much deeper in the human spirit than a souvenir of a vainglorious triumph over a defeated animal. In the fourteenth-century *Taymouth Hours* there are thirty scenes with women in hunting rituals that are depicted with the removal of the beast's entrails and the head of the animal set on a pole, announcing the kill. Human heads may well have been stuck on posts or kicked about to shame and taunt the losers in battle, but that may be only the degraded vestige of a much more positive tradition. The heads are set in niches in Gaulish monuments in southern France. Consider all the human heads and busts that are a principal residue of the celebration of great men and women. Even in books we illustrate them as heads, as though it preserves a profound human assumption about heads, namely that the spirit resides mostly there.

"Sport killing" seems opposed to ethnic tradition. Old photographs of white hunters with piles of dead animals—who would consider defending such "slaughter"? As greed or brutality it may overlay an older impulse of which it is a perversion. The trophy display descends from the tradition of laying out the dead animals, as well as the belief that they continue to be consciously present. In an unpublished manuscript a modern hunter, C. H. D. Clarke, writes: "The Mexican Indian shamanic deer hunt is as much pure sport as mine, and the parallels between its rituals, where the dead game is laid out in state, and those of European hunts, where the horns sound the 'Sorbiati,' or 'tears of the stag,' over the dead quarry, are beyond coincidence."

10. A. H. Ismail and L. E. Trachtman, "Jogging the Imagination," *Psychology Today*, March 1973, pp. 78–82.

11. Hyemeyohsts Storm, *Seven Arrows* (New York: Harper & Row, 1972), p. 344.

12. Gary Snyder, "Four Changes," in *Turtle Island* (New York: New Directions, 1974), pp. 96–97.

13. Jim Cheney, "The Waters of Separation," *Journal of Feminist Studies in Religion* 6(1) (Spring 1990): 41–60.

14. Ernest Becker, *The Denial of Death* (New York: Free Press, 1974).

15. Erik H. Erikson, "Youth: Fidelity and Diversity," *Daedalus* 81(1) (1962): 6.

16. James Mischke, "Legends of Warriors and Misogynists: Roots of the Contemporary Navajo Cultural Quandary in History, Mythos, and Social Research," unpublished manuscript, 1995.

17. Daniel Goleman, "Fear of Death Intensifies Moral Code, Scientists Find," *New York Times*, December 1989, pp. C-1 and C-11.

18. See, for instance, Robert W. Loftin, "The Morality of Hunting," *Environmental Ethics* 6(3) (1984): 241–250.

19. Alexander Marshack, *The Roots of Civilization* (New York: McGraw-Hill, 1972).

20. Robert Ridington, "Beaver Dreaming and Singing," in David M. Guss, ed., *The Language of the Birds* (San Francisco: North Point, 1985), p. 52.

21. N. K. Sandars, *Prehistoric Art of Europe*, 2nd ed. (London: Pelican, 1985), pp. 26–27.

22. Robert Ardrey, the gifted author of *The Territorial Imperative* (New York: Harper & Row, 1968), though he was right about much else, was wrong about territoriality as a human condition. Exclusive use of places is rare among foraging peoples. It appears in history as a concomitant of war and aggression among agricultural groups and city-states.

23. C. L. Rawlings in *Sky's Witness* (p. 236) says: "Walking, along with grasping and talking, is a human birthright, but few of us do it well. Listen to the way your foot meets the earth. Is there noise? Set aside the lore about walking silently, since we are not sneaking up on sabertooths, and consider the physics. We walk by rolling the body's weight forward and falling slightly, then swinging a leg out to arrest the fall, placing a foot on the ground. Any sound, slapping, scuffing, sliding, thumping, means that there is energy being wasted. Your foot should descend softly and kiss the earth. Scuffing means your foot is being dragged. If it slides back as you launch, you may be taking overlong strides, leaning forward too much. Make clean, sharp tracks."

24. Raymond Chipeniuk, "Childhood Foraging as a Means of Acquiring Competent Human Cognition about Biodiversity," *Environment and Behavior* 27(4) (July 1995): 490–512.

25. Raymond Chipeniuk, "The Sense of Naturalness: A Transcultural Approach to Environmental Citizenship," unpublished manuscript, 1993.

26. Ibid., pp. 19–21.

27. Konrad Lorenz and Paul Leyhausen, *Motivation of Human and Animal Behavior: An Ethological View* (New York: Van Nostrand, 1973).

28. I remember how dramatic Wendell Wilkie's book, *One World*, seemed to us when I was a freshman in high school. We have been inundated with appeals to a single unity ever since: universal humanity, the planet, the earth, the biosphere. The rhetoric of oneness leads to distinctions and oppositions: the bioregion, the ecosystem, the culture, and of course, our special baby, the individual. And these special categories in turn give rise to language that erases the differences: multiculturalism, United Nationals, international law, "think globally, live locally." Much of our thought is set up in these polar oppositions with their middle-ground solutions. Perhaps we have asked the wrong question and sought impossible solutions. Lévi-Strauss says: "It would not be enough to absorb particular humanities into a general one." We must proceed, rather, to "the reintegration of culture in nature." Their opposition, he says, is merely methodological.

Bibliography

Abrams, H. Leon, Jr. "The Preference for Animal Proteins and Fat: A Cross-Cultural Survey." In Marvin Harris and Eric B. Ross, eds., *Food and Evolution.* Philadelphia: Temple University Press, 1985.

Adams, Carol. *The Sexual Politics of Meat.* New York: Continuum, 1990.

Allen, Robert. "Food for Thought." *The Ecologist,* January 1975.

Altherr, Thomas L., and John F. Reiger. "Academic Historians and Hunting: A Call for More and Better Scholarship." *Environmental History Review* 19(3) (Fall 1995).

Angus, S. *The Mystery Religions.* New York: Dover, 1975. First published in 1925.

Ardrey, Robert. *The Hunting Hypothesis.* New York: Atheneum, 1976.

——. *The Territorial Imperative: A Personal Inquiry into the Animal Origins of Property and Nature.* New York: Harper & Row, 1968.

Arnheim, Rudolf. *Towards a Psychology of Art.* Berkeley: University of California Press, 1954.

Arrowsmith, William Ayres. "Hybris and Sophrosyne." *Dartmouth Alumni Magazine,* July 1970.

Bailey, Liberty Hyde. *The Holy Earth.* New York: Scribner's, 1915.

Becker, Ernest. *The Denial of Death.* New York: The Free Press, 1974.

Berman, Morris. "The Roots of Reality." (Review of Humberto Maturana and Francisco Varelas, *The Tree of Knowledge.*) *Journal of Humanistic Psychology* 23(2) (1989): 277–284.

Berry, John W., and Robert C. Annis. "Ecology, Culture, and Psychological Differentiation." *International Journal of Psychology* 9(3) (1974): 173–193.

Berry, R. J. "The Genetical Implications of Domestication in Animals." In Peter J. Ucko and G. W. Dimbleby, eds., *The Domestication and Exploitation of Plants and Animals.* Chicago: Aldine, 1969.

Biesele, Megan. "Aspects of !Kung Folklore." In Richard E. Lee and Irven

DeVore, eds., *Kalahari Hunters and Gatherers.* Cambridge, Mass.: Harvard University Press, 1976.

Bird-David, Nurit. "The Giving Environment." *Current Anthropology* 31(2) (April 1990): 189–196.

Bloch, Maurice, and Jonathan Parry. *Death and the Regeneration of Life.* New York: Cambridge University Press, 1982.

Blumenschine, Robert J., and John A. Cavallo. "Scavenging and Human Evolution." *Scientific American,* October 1992, pp. 90–91.

Boas, George. *Essays on Primitivism and Related Ideas in the Middle Ages.* Baltimore: Johns Hopkins University Press, 1948.

Bookchin, Murray. *The Rise of Urbanization and the Decline of Citizenship.* San Francisco: Sierra Club Books, 1987.

Bronowski, Jacob. *The Ascent of Man.* 1st American ed. Boston: Little, Brown, 1973.

Brotherston, Gordon. "Andean Pastoralism and Inca Ideology." In Juliet Clutton-Brock, ed., *The Walking Larder.* London: Unwin Hyman, 1990.

Brown, Cecil H. H. "Mode of Subsistence and Folk Biological Taxonomy." *Current Anthropology* 26(1) (1985): 43–53.

Brown, Norman O. *Life against Death: The Psychoanalytical Meaning of History.* Middletown, Conn.: Wesleyan University Press, 1959.

Butterworth, C. A. S. *Some Traces of the Pre-Olympian World in Greek Literature and Myth.* Berlin: De Gruyter, 1966.

Butzer, Karl W. *Environment and Archeology: An Ecological Approach to Prehistory.* 2nd ed. Chicago: Aldine Atherton, 1971.

Calder, Nigel. *Eden Was No Garden.* New York: Holt, 1967.

Campbell, Joseph. *The Masks of God.* Vol. 1. New York: Viking Press, 1959.

——. "The Way of the Seeded Earth, Part I." *Historical Atlas of World Mythology,* Vol. 2. New York: Harper & Row, 1988.

Carse, James. *Finite and Infinite Games: A Vision of Life as Play and Possibilities.* New York: Ballantine, 1986.

Cartmill, Matt. *A View to Death in the Morning.* Cambridge, Mass.: Harvard University Press, 1993.

Cavalli-Sforza, L. L. "The Transition to Agriculture and Some of Its Consequences." In Donald J. Ortner, ed., *How Humans Adapt.* Washington, D.C.: Smithsonian, 1983.

Chance, Michael R. A., ed. *Social Fabric of the Mind.* London: Erlbaum, 1988.

Chaplin, Raymond E. "The Use of Non-Morphological Criteria in the Study of Animal Domestication from Bones Found on Archaeological Sites." In Peter J. Ucko and G. W. Dimbleby, eds., *The Domestication and Exploitation of Plants and Animals.* Chicago: Aldine, 1969.

Cheney, Jim. "The Waters of Separation." *Journal of Feminist Studies in Religion* 6(1) (Spring 1990): 41–63.

Chipeniuk, Raymond. "Childhood Foraging as a Means of Acquiring Competent Human Cognition about Biodiversity." *Environment and Behavior* 27(4) (July 1995): 490–512.

———. "The Sense of Naturalness: A Transcultural Approach to Environmental Citizenship." Unpublished manuscript, 1993.

Chowka, Peter B. "The Original Mind of Gary Snyder." *East-West,* June 1977.

Cixous, Helene, and Catherine Clement. *The Newly Born Woman.* Minneapolis: University of Minnesota Press, 1986.

Clarke, C. H. D. "Venator the Hunter." Unpublished manuscript, n.d.

Clifton, Charles. "Hunter's Eucharist." *Gnosis,* Fall 1993.

Clutton-Brock, Juliet, ed. *The Walking Larder.* London: Unwin Hyman, 1990.

Cobb, Edith. *The Ecology of Imagination in Childhood.* New York: Columbia University Press, 1977.

Cobb, John B., Jr. *The Structure of Christian Existence.* Philadelphia: Westminster, 1967.

Cranston, B. A. L. "Animal Husbandry: The Evidence from Ethnography." In Peter J. Ucko and G. W. Dimbleby, eds., *The Domestication and Exploitation of Plants and Animals.* Chicago: Aldine, 1969.

Crownfield, David. "The Curse of Abel: An Essay in Biblical Ecology." *North American Review* 258(2) (1973): 58–63.

Darling, Frank. *Wilderness and Plenty.* Reith Lecture. London: BBC, 1970.

Diamond, Stanley. *In Search of the Primitive.* New Brunswick: Transaction, 1974.

Draper, Patricia. "Social and Economic Constraints on Child Life among the !Kung." In Richard B. Lee and Irven DeVore, eds., *Kalahari Hunters and Gatherers.* Cambridge, Mass.: Harvard University Press, 1976.

Dubos, Rene. "Environmental Determinants of Human Life." In David C. Glass, ed., *Environmental Influences.* New York: Rockefeller University Press, 1968.

Dupre, Wilhelm, ed. *Religion in Primitive Cultures: A Study in Ethnography.* The Hague: Mouton, 1975.

Durant, Will. "A Last Testament to Youth." *The Columbia Dispatch Magazine,* 8 February 1970.

Eaton, Stanley Boyd, and Marjorie Shostak. "Fat Tooth Blues." *Natural History* 95(6) (July 1986).

Eaton, Stanley Boyd, Marjorie Shostak, and Melvin Konner. *The Paleolithic Prescription: A Program of Diet & Exercise and a Design for Living.* New York: Harper & Row, 1988.

Edgerton, Robert. *The Individual in Cultural Adaptation.* Los Angeles: University of California, 1971.

———. *Sick Societies: Challenging the Myth of Primitive Harmony.* New York: Free Press, 1992.

Eibl-Eibesfeldt, Irenaus. *Love and Hate: The Natural History of Behavior Patterns.* New York: Aldine de Gruyter, 1996.

Eiseley, Loren. "Man of the Future." *The Immense Journey.* New York: Random House, 1959.

Erikson, Erik H. *The Life Cycle Completed.* New York: Norton, 1985.

——. "Youth: Fidelity and Diversity." *Daedalus* 81(1) (1962): 6.

Erwin, Robert. *The Great Language Panic.* Athens: University of Georgia Press, 1990.

Fernandez, James. "Review of *Meditations on Hunting* by Ortega and Gassett." *American Anthropologist* 76(4) (1974): 868–869.

Flannery, Kent. "Origins and Ecological Effects of Early Domestication in Iran and the Near East." In Peter J. Ucko and G. W. Dimbleby, eds., *The Domestication and Exploitation of Plants and Animals.* Chicago: Aldine, 1969.

Foster, Hal. "Signs Taken for Wonders." *Art in America.* June 1986, pp. 80–139.

Fowles, John. *A Maggot.* Boston: Little, Brown, 1985.

Freeman, Derek. Letter. *Current Anthropology* 14(4) (1973): 379.

Fuentes, Carlos. *Christopher Unborn.* New York: Farrar, Straus Giroux, 1989.

Gadacz, Renez R. "Montagnais Hunting Dynamics in Historicoecological Perspective." *Anthropologica* 17(2) (1975): 149–168.

Galaty, John G. "Cattle and Cognition: Aspects of Masai Practical Reasoning." In Juliet Clutton-Brock, ed., *The Walking Larder.* London: Unwin Hyman, 1990)

Geist, Valerius. "Did Large Predators Keep Humans Out of North America?" In Juliet Clutton-Brock, ed., *The Walking Larder.* London: Unwin Hyman, 1990.

Gimbutus, Marija. *Goddesses and Gods of Old Europe.* Berkeley: University of California Press, 1982.

Gleick, James. *Chaos: Making a New Science.* New York: Viking, 1987.

Goleman, Daniel. "Fear of Death Intensifies Moral Code, Scientists Find." *New York Times,* 5 December 1989, pp. C-1 and C-11.

Gordon, Cyrus H. *The Common Background of Greek and Hebrew Civilizations.* New York: Norton, 1965.

Gowdy, John. "The Bioethics of Hunting and Gathering Societies." *Social Economy* 50(2) (Summer 1992): 130–149.

Grayson, Donald. "Pleistocene Avifauna and the Overkill Hypothesis." *Science* 18 February 1977, pp. 691–692.

Guss, David M., ed. *The Language of the Birds.* San Francisco: North Point, 1985.

Hallowell , A. Irving. "Self, Society, and Culture in Phylogenetic Perspective." In Sol Tax, ed., *The Evolution of Man.* Chicago: University of Chicago Press, 1960.

Harding, Robert S. O. "An Order of Omnivores: Nonhuman Primate Diets in the Wild." In Robert S. O. Harding and Geza Teleki, eds., *Omnivorous Primates: Gathering and Hunting Diets in the World.* New York: Columbia University Press, 1981.

Harris, David. "Agricultural Systems, Ecosystems, and the Origin of Agriculture." In Peter J. Ucko and G. W. Dimbleby, eds., *The Domestication and Exploitation of Plants and Animals.* Chicago: Aldine, 1969.

Harris, Marvin. *The Sacred Cow and the Abominable Pig.* New York: Simon & Schuster, 1987.

Harris, Marvin, and Eric B. Ross, eds. *Food and Evolution.* Philadelphia: Temple University Press, 1985.

Harrison, Jane Ellen. *Themis.* New York: University Books, 1962.

Herre, Wolfe, and Manfred Röhrs. "Zoological Considerations on the Origins of Farming and Domestication." In Charles A. Reed, ed., *The Origins of Agriculture.* The Hague: Mouton, 1977.

Hillman, James. "Senex and Puer." *Puer Papers.* Dallas: Spring Publications, 1979.

Hole, Frank and Kent Flannery. "The Prehistory of Southwestern Iran: A Preliminary Report." *Proceedings of the Prehistoric Society* 33 (1963): 201.

Howard, Jane. "All Happy Clans Are Alike." *Atlantic,* May 1978.

Hutchins, Robert. Preface to Mortimer J. Adler's Hundred Great Books Series, *The Great Ideas.* Chicago: Encyclopedia Britannica, 1952.

Huxley, Aldous. "Mother." In *Tomorrow and Tomorrow and Tomorrow.* New York: Harper, 1952.

Hyde, Lewis. *The Gift.* New York: Random House, 1979.

Illich, Ivan. *Gender.* New York: Pantheon, 1982.

Iltis, Hugh. "Flowers and Human Ecology." *New Movement in the Study and Teaching of Biology.* London: Maurice Temple Smith, 1974.

Ingold, Timm. *The Appropriation of Nature: Essays on Human Ecology and Social Relations.* Iowa City: University of Iowa Press, 1987.

Isaac, Glynn L. I., and Diana C. Crader. "To What Extent Were Early Hominids Carnivorous? An Archaeological Perspective." In Robert S. O. Harding and Geza Teleki, eds., *Omnivorous Primates: Gathering and Hunting Diets in the World.* New York: Columbia University Press, 1981.

Ismail, A. H., and L. E. Trachtman. "Jogging the Imagination." *Psychology Today,* March 1973, pp. 78–82.

Jacobson, Esther. *The Deer Goddess of Ancient Siberia: A Study in the Ecology of Belief.* New York: Brill, 1993.

James, E. O. *The Worship of the Sky-God.* London: University of London Press, 1963.

Jerison, Harry J. *The Evolution of the Brain and Intelligence.* New York: Academic Press, 1973.

Jochim, Michael A. *Hunter-Gatherer Subsistence and Settlement: A Predictive Model.* New York: Academic Press, 1976.

Johnson, Allen W., and Timothy Earle. *The Evolution of Human Societies from Foraging Group to Agrarian State.* Palo Alto: Stanford University Press, 1987.

Jones, Kevin T. "Hunting and Scavenging by Early Hominids: A Study in Archaeological Method and Theory." Ph.D. thesis, University of Utah, 1984.

Keesing, Roger M. "Paradigms Lost: The New Ethnography and New Linguistics." *Southwest Journal of Anthropology* 28 (1972): 299–332.

Kent , Susan. *Farmers as Hunters: The Implications of Sedentism.* New York: Cambridge University Press, 1989.

Knauft, Bruch M. "Violence and Sociality in Human Evolution." *Current Anthropology* 32 (1991): 391–428.

Kolata, Gina Bari. "!Kung Hunter-Gatherers: Feminism, Diet, and Birth Control." *Science* 195(4276) (28 January 1977): 382–383.

Konner, Melvin J. "Maternal Care, Infant Behavior and Development among the !Kung." In Richard B. Lee and Irven DeVore, eds., *Kalahari Hunters and Gatherers.* Cambridge, Mass.: Harvard University Press, 1976.

——. *The Tangled Wing: Biological Constraints on the Human Spirit.* New York: Holt, 1982.

Kristeva, Julia. *Tales of Love.* New York: Columbia University Press, 1987.

LaChapelle, Dolores. *D. H. Lawrence, Future Primitive.* College Station: University of North Texas Press, 1996.

Lasch, Christopher. *The True and Only Heaven.* New York: Norton, 1991.

Lavery, David. *Late for the Sky.* Carbondale: Southern Illinois University Press, 1992.

Lawrence, Elizabeth. *Hoofbeats and Society.* Bloomington: University of Indiana Press, 1985.

Layard, John. "The Incest Taboo and the Virgin Archetype." In Joanne Stroud and Gail Thomas, eds., *Images of the Untouched.* Dallas: Spring Publications, 1992.

Lee, Dorothy. "Lineal and Non-Lineal Codifications of Reality." *Psychosomatic Medicine* 12(2) (1950): 89–97.

Lee, Richard Borshay. *The !Kung San: Men, Women, and Work in Foraging Society.* New York: Cambridge University Press, 1979.

——. *Kalahari Hunters and Gatherers: Studies in the !Kung San and Their Neighbors.* Cambridge, Mass.: Harvard University Press, 1976.

Lee, Richard B., and Irven DeVore, eds. *Man the Hunter.* Chicago: Aldine, 1968.

Lévi-Strauss, Claude. *The Savage Mind.* Chicago: University of Chicago Press, 1966.

Levy, G. Rachel. *Religious Conceptions of the Stone Age and Their Influences on European Thought.* New York: Harper & Row, 1963.

Lewin, Bertram. *The Image and the Past.* New York: International Universities Press, 1968.

Lincoln, Bruce. *Priests, Warriors, and Cattle: A Study of the Ecology of Religion.* Berkeley: University of California Press, 1981.

Loftin, Robert W. "The Morality of Hunting." *Environmental Ethics* 6(3) (1984): 241–250.

Lopreato, J. "How Would You Like to Be a Peasant?" *Human Organization* 24 (1965): 4.

Lott, Dale F., and Ben L. Hart. "Aggressive Domination of Cattle by Fulani Herdsmen and Its Relation to Aggression in Fulani Culture and Personality." *Ethos* 5 (1977): 174–186.

Lotz, Pat, and Jim Lotz, eds. "Pilot, Not Commander: Essays in Memory of Diamond Jenness." *Anthropologia* n.s. (13) (1971).

Lowenthal, David. "Is Wilderness 'Paradise Enow'?" *Columbia University Forum* 7(2) (Spring 1968): 34–40.

Mann, Alan E. "Diet and Human Evolution." In Robert S. O. Harding and Geza Teleki, eds., *Omnivorous Primates.* New York: Columbia University Press, 1981.

Marano, Hara Estroff. "Biology Is One Key to the Bonding of Mothers and Babies." *Smithsonian,* February 1981, pp. 60–68.

Marcuse, Herbert. *One Dimensional Man.* Boston: Beacon, 1964.

Martin, Calvin. *Keepers of the Game.* Los Angeles: University of California Press, 1978.

Marx, Emanuel. "The Ecology and Politics of Nomadic Pastoralists in the Middle East." In Wolfgang Weissleder, ed., *The Normadic Alternative.* The Hague: Mouton, 1978.

Matthiessen, Peter. "Survival of the Hunter." *New Yorker,* 24 April 1995, pp. 67–77.

McLuhan, Marshall, and Harley Parker. *Through the Vanishing Point: Space in Poetry and Painting.* New York: Harper & Row, 1968.

Mischke, James. "Legends of Warriors and Misogynists: Roots of the Contemporary Navajo Cultural Quandary in History, Mythos, and Social Research." Unpublished manuscript, 1995.

Modell, Arnold. *Object Love and Reality.* New York: International Universities Press, 1968.

Morell, Virginia. "Stone Age Menagerie." *Audubon,* May–June 1995.

Muller, Herbert J. *The Uses of the Past.* New York: Oxford University Press, 1952.

Mumford, Lewis. "The First Megamachine." *Diogenes,* Fall 1966, pp. 1–5.

——. *The Myth of the Machine.* New York: Harcourt Brace, 1966.

Murdock, G. P. *Ethnographic Atlas for New World Societies.* Pittsburgh: University of Pittsburgh Press, 1967.

Myer, Fred. *Pintupi Country, Pintupi Self: Sentiment, Place, and Politics Among Desert Aborigines.* Washington, D.C.: Smithsonian, 1986.

Nabokov, Peter. *Indian Running.* Santa Barbara: Capra, 1981.

Neel, James V. "Lessons from a Primitive People." *Science* 170(3960) (20 November 1970): 815–822.

Nelson, Richard. *Make Prayers to the Raven: A Koyukan View of the Northern Forest.* Chicago: University of Chicago Press, 1983.

North, Douglas C., and Robert Paul Thomas. "The First Economic Revolution." *The Economic History Review* 30(2) (1970): 229–241.

O'Brien, Bogert. "Inuit Ways and the Transformation of Canadian Theology." Unpublished manuscript, 1979.

Ong, Walter J. "World as View and World as Event." *American Anthropologist* 71(4) (1969): 634–647.

Ortega y Gasset, José. *Meditations on Hunting.* Translated by Howard B. Wescott. Introduction by Paul Shepard. New York: Scribner's, 1972.

Ortner, Donald J., ed. *How Humans Adapt.* Washington, D.C.: Smithsonian, 1983.

Pagdon, Anthony. *The Fall of Natural Man.* New York: Cambridge University Press, 1982.

Paz, Octavio. *The Other Mexico: Critique of the Pyramid.* New York: Grove Press, 1972.

Peters, Charles R. "Toward an Ecological Model of African Plio-Pleistocene Hominid Adaptations." *American Anthropologist* 81(2) (1979): 261–278.

Potter, Jack M. *Peasant Society.* Boston: Little, Brown, 1967.

Rawlins, C. L. *Sky's Witness.* New York: Holt, 1993.

Riches, David. "Hunting, Herding and Potlaching: Towards a Sociological Account of Prestige." *Man* 19(2) (1984): 234–257.

Ridington, Robert. "Beaver Dreaming and Singing." In David M. Guss, ed., *The Languages of the Birds.* San Francisco: North Point, 1985.

Robson, J. R. K., ed. *Food, Ecology and Culture: Readings in the Anthropology of Dietary Practice.* New York: Gordon & Breach, 1980.

Roheim, Geza. *The Children of the Desert: The Western Tribes of Central Australia.* New York: Basic Books, 1974.

Rolston, Holmes, III. "Values Gone Wild." *Inquiry* 26(2) (1983): 181–207.

Roseman, Marina. "The Social Structure of Sound: The Temiar of Peninsular Malaysia." *Symposium of Comparative Musicology.* Proceedings of the Society for Ethnomusicality, 29th annual meeting, University of California, Los Angeles, 18–21 October, 1984.

Ross, Eric B. "An Overview of Trends in Dietary Variation from Hunter-Gatherer to Modern Capitalist Societies." In Marvin Harris and Eric B. Ross, eds., *Food and Evolution.* Philadelphia: Temple University Press, 1985.

Rutkowska, Julia. "Does the Phylogeny of Conceptual Development Increase Our Understanding of Concepts or of Development?" In George Butterworth et al., eds., *Evolution and the Development of Psychology.* New York: St. Martin's Press, 1985.

Saint-Exupéry, Antoine de. *The Wisdom of the Sands.* New York: Harcourt Brace, 1950.

Sandars, N. K. *Prehistoric Art of Europe.* 2nd ed. London: Pelican, 1985.

Sarno, Louis. *Bayaka: The Extraordinary Music of the Babenzélé Pygmies and Sounds of Their Forest Home.* With producer's notes and compact disc by Bernie Krause. Roslyn, N.Y.: Ellipsis Arts, 1995.

Schneidau, Herbert. *Sacred Discontent: The Bible and Western Tradition.* Baton Rouge: Louisiana State University Press, 1976.

Schneider, Jane. "Of Vigilance and Virgins: Honor, Shame and Access to Resources in Mediterranean Societies." *Ethnology* 10(6) (1971): 1–24.

Schwabe, Calvin W. "The Most Intense Man–Animal Bond." Unpublished manuscript, n.d.

Searles, Harold F. *The Nonhuman Environment.* New York: International Universities Press, 1960.

Service, Elman R. *The Hunters.* 2nd ed. Englewood Cliffs: Prentice-Hall, 1979.

Shepard, Paul. *Man in the Landscape: A Historic View of the Esthetics of Nature.* College Station: Texas A&M University Press, 1991.

——. *Nature and Madness.* San Francisco: Sierra Club Books, 1982.

——. *The Others: How Animals Made Us Human.* Washington, D.C.: Island Press, 1995.

——. "A Post-Historic Primitivism." In Max Oelschlaeger, ed., *The Wilderness Condition: Essays on Environment and Civilization.* San Francisco: Sierra Club Books, 1992.

——. *The Sacred Paw: The Bear in Nature, Myth and Literature.* New York: Viking, 1985.

——. *The Tender Carnivore and the Sacred Game.* New York: Scribner's, 1973.

——. *Thinking Animals.* New York: Viking, 1978.

——. "Wilderness Is Where My Genome Lives." *Whole Terrain* 4 (1995–1996): 12–16.

Shostak, Marjorie. "A !Kung Woman's Memories of Childhood." In Richard B. Lee and Irven DeVore, eds., *Kalahari Hunters and Gatherers.* Cambridge, Mass.: Harvard University Press, 1976.

Shostak, Marjorie, and Melvin Konner. *The Paleolithic Prescription: A Program of Diet & Exercise and a Design for Living.* New York: Harper & Row, 1988.

Simoons, Frederick J. "The Determinants of Dairying and Milk Use in the Old World: Ecological, Physiological, and Cultural." In J. R. K. Robson, ed., *Food, Ecology and Culture.* New York: Gordon & Breach, 1980.

Siskind, Janet. *To Hunt in the Morning.* New York: Oxford University Press, 1973.

Slater, Philip. *Earthwalk.* Garden City: Doubleday, 1974.

Snyder, Gary. *Turtle Island.* New York: New Dictionary, 1974.

Sontag, Susan. *On Photography.* New York: Dell, 1973.

Sorkin, Michael. "See You in Disneyland." In Michael Sorkin, ed., *Variations on a Theme Park: The New American City and the End of Public Space.* New York: Hill & Wang, 1992.

Spurway, Helen. "The Causes of Domestication." *Journal of Genetics,* 5(3) (1955): 325–327.

Stanner, W. E. H. *White Man Got No Dreaming: Essays 1938–1973.* Canberra: Australian National University Press, 1979.

Storm, Hyemeyohsts. *Seven Arrows.* New York: Harper & Row, 1972.

Strum, Shirley C. "Processes and Products of Change: Baboon Predatory Behavior at Gigil, Kenya." In Robert S. O. Harding and Geza Teleki, eds., *Omnivorous Primates.* New York: Columbia University Press, 1981.

Swan, James A. *In Defense of Hunting.* San Francisco: Harper, 1995.

Tax, Sol, ed. *The Evolution of Man.* Vol. 2. Chicago: University of Chicago Press, 1960.

Teleki, Geza. "The Omnivorous Diet and Eclectic Feeding Habits of

Chimpanzees in Gombe National Park, Tanzania." In Robert S. O. Harding and Geza Teleki, eds., *Omnivorous Primates.* New York: Columbia University Press, 1981.

Trilling, Lionell. *Beyond Culture: Essays on Literature and Learning.* New York: Viking Press, 1955.

Turnbull, Colin. *The Human Cycle.* New York: Simon & Schuster, 1983.

Tylor, Gordon Rattray. *Rethink: A Paraprimitive Solution.* New York: Dutton, 1973.

Ucko, Peter J., and G. W. Dimbleby. "Introduction." In *The Domestication and Exploitation of Plants and Animals.* Chicago: Aldine, 1969.

Urton, Gary, ed. *Animal Myths and Metaphors in South America.* Salt Lake City: University of Utah Press, 1985.

van der Post, Laurens. *Heart of the Hunter.* New York: Harcourt Brace Jovanovich, 1980.

Washburn, Sherwood L., ed. *The Social Life of Early Man.* New York: Wenner-Gren Foundation, 1961.

Watcher, D. N., and N. Kretchner, eds. *Nutrition and Evolution.* New York: Masson, 1981.

Weissleder, Wolfgang. *The Nomadic Alternative.* The Hague: Mouton, 1978.

White, Lynn. *Medieval Technology and Social Change.* New York: Oxford University Press, 1970.

Wood, J. W., et al. "The Osteological Paradox." *Current Anthropology* 33(4) (1992): 343–370.

Woodburn, James. "An Introduction to Hadza Ecology." In Richard B. Lee and Irven DeVore, eds., *Man the Hunter.* Chicago: Aldine, 1968.

Zink, Nelson, and Stephen Parks. "Nightwalking: Exploring the Dark with Peripheral Vision." *Whole Earth Review,* Fall 1991, pp. 4–9.

Zvelebil, Marek. "Postglacial Foraging in the Forests of Europe." *Scientific American,* May 1986, p. 104.

Index